Volney Rattan

West Coast Botany

An Analytical key to the Flora of the Pacific Coast in Which are Described Over

Eighteen Hundred Species of Flowering Plants Growing West of the Sierra Nevada

and Cascade Crests, from San Diego to Puget Sound

Volney Rattan

West Coast Botany
*An Analytical key to the Flora of the Pacific Coast in Which are Described Over Eighteen
Hundred Species of Flowering Plants Growing West of the Sierra Nevada and Cascade Crests,
from San Diego to Puget Sound*

ISBN/EAN: 9783337267070

Printed in Europe, USA, Canada, Australia, Japan

Cover: Foto ©berggeist007 / pixelio.de

More available books at **www.hansebooks.com**

WEST COAST BOTANY,

AN

ANALYTICAL KEY

TO THE

FLORA OF THE PACIFIC COAST,

IN WHICH ARE DESCRIBED OVER

EIGHTEEN HUNDRED SPECIES OF FLOWERING PLANTS GROWING
WEST OF THE SIERRA NEVADA AND CASCADE CRESTS,
FROM SAN DIEGO TO PUGET SOUND.

BY

VOLNEY RATTAN,

TEACHER OF BOTANY IN THE STATE NORMAL SCHOOL, SAN JOSE, CALIFORNIA.

SAN FRANCISCO,
THE WHITAKER & RAY CO.,
1898.

PREFACE.

The skeleton of this book has for eleven years formed a supplement to the "California Flora," which describes only the plants of the coast region between Monterey and Ukiah. Since it is not practicable to fill out this skeleton in the way originally intended, it has been put into the improved form here described. The "Flora" of the old manual has been replaced by descriptions of all the orders whose species on this coast have conspicuous flowers. This part of the book also contains descriptions of near two hundred and fifty species which are mostly new, and over fifty generic names which in Greene's "Botany of the San Francisco Bay Region" displace names used in this book. These synonyms will be very helpful to those who use the former manual with this or other floras. A complete glossary of the botanical terms and specific names found in this book, and a glossary of generic names in connection with the index will materially aid students. An analytical key leads the student to a description of the order to which the plant in hand belongs. At the close of that description he is referred to the page of the second part where keys lead to the genus and species. Returning to the first part, the new matter there is consulted before making a final decision. This seemingly awkward prominence of addenda is perhaps advantageous to the student, who is thus led to realize the progress of botanical work. Moreover there is encouragement in the thought that the discovery of so many new plants in the ten years just ended proves that there are species yet undescribed which sharp-eyed seekers may find.

Since the descriptions here given are for the most part abbreviations of the originals, their shortcomings should be charged to the compiler of this book. In some cases, particularly in difficult genera, new species have not been given. Many of Prof. Greene's new species are placed under generic names which he does not approve. Most of these, however, were described by him under the discarded names, and it is proper that his name should follow that of the species as author. In the other cases his name, according to present usage, should appear in parenthesis. Since this is a matter of little importance to beginners and one difficult to manage it has not been attended to.

Because of the unsettled condition of plant names the present time is unfavorable for the preparation of a flora of any country. More than ever before systematic botanists are investigating the history of names, and, like other historians, they do not agree. There are therefore added to the ever present questions concerning the limitations of genera and species, questions concerning the priority of names. The former never will be settled, and authorities are not likely to agree upon the latter for some years to come. Meanwhile we must learn several names for each of a score or more of the plants we yearly greet in our country rambles. For example: In the collections of plants made in the United States last year the shrub commonly known as Nine-Bark doubtless bears five different names. Those using "Gray's Manual" or "Wood's Class Book" have labeled it *Spiræa opulifolia;* according to "Bergen's Botany" and "Behr's Flora" it is *Neillia opulifolia;* in the "Key to West Coast Botany" it is *Physocarpus opulifolia;* in Greene's "Flora Franciscana" it is *Neillia capitata,* and in the same author's "Botany of the Bay Region" it is called *Opulaster capitatus.* Five plants—so common that they may be found on one hillside—will, by those who use the "Bay Region Botany," be given each a separate generic name, yet most botanists call them all *Gilias.* A common wild cherry is *Prunus emarginata* in "Behr's Flora," *Cerasus emarginata* in "Bay Region Botany," and *Cerasus Californica* in "Flora Franciscana." Some idea of the number of plants known by more than one name may be gained from the fact that over two hundred of the thirteen hundred species described in the "Bay Region Botany" appear under generic and sometimes specific names different from those given them in the "Botany of the Geological Survey." But it must be remembered that, even in its present chaotic condition, botanical nomenclature is incomparably better than that of so-called common names. Most of our noticeable native plants are each known by a dozen or more local names.   V. R.

San Jose, Feb. 8, 1898.

SYSTEMATIC BOTANY.

NAMES OF PLANTS: CLASSIFICATION.

In a general way we designate the objects around us by single names. We speak of a stone, a wolf, or a pine; but to distinguish the kinds we naturally use two names, as lime stone, sand stone; grey wolf, prairie wolf; nut pine, yellow pine, etc. This is one step in classification, and the only one commonly taken. This natural plan of double names was adopted by the great naturalist, Linnæus, who gave names to most European plants, as well as to many of this continent. He wisely gave the Latin form to his names, since that language (being the base of most languages spoken in civilized countries) is the natural source of cosmopolitan names—those truly common to all people. Botanical names, then, differ from so-called common names principally in form, and they have these decided advantages: they more exactly represent the relations between kinds of plants, and they are names that are common to people of all languages. In short, they are the true *common names*.

It is not true that botanical names are harder than local names. The most common of our ornamental plants are well known by their scientific names. No one thinks of calling the following botanical names hard: Geranium, Aster, Verbena, Petunia, Portulaca, Crocus, Phlox, Fuchsia, Iris, Magnolia, Oxalis, Azalea, Dahlia, Lobelia, Arnica, etc. Most people talk familiarly of Camellias, Callas, Begonias, Acacias, etc.; while our beautiful California plants, Clarkia, Collinsia, Eschscholtzia, Nemophila, etc., are well known by their proper names—at least, in other countries.

Generic names correspond to the second parts of the compound common names, as oak, pine, rose, etc. Some of these are the old Greek or Latin names of the plant. Most generic names are either derived from Greek or Latin words descriptive of some peculiarity of the plant, or they are commemorative of some botanist, as Thysanocarpus, from Greek words meaning fringe and pod; Kelloggia, in honor of Dr. A. Kellogg, a veteran botanist of this coast. Sometimes genera are named in honor of those who are not botanists, as Fremontia, Hollisteria, Stanfordia, etc.

It will be seen that in the examples given a generic name in honor of a man is formed by adding "ia" to his name. Sometimes "a" only is added, as Bolandra.

Specific names correspond to the first part of common names, but are written after the generic names. Thus Oregon Oxalis is labeled *Oxalis Oregona*. Most specific names are descriptive, as *Gilia tricolor*, Tricolored Gilia. Frequently a species is named for the discoverer, as *Eriogonum Nortoni*, Norton's Eriogonum; or in honor of some one,

cabbage, radish. But you can do nothing with double flowers. A sweet pea could be made to tell its proper or generic name in this way:

The sepals and petals together more than six, and petals not all united, brings us again to **Division 1.** This time "**A. STAMENS MORE THAN TEN**" is wrong. We take "**B. STAMENS TEN OR LESS.**" "Ovary or ovaries superior," etc., is right, but * "*pistils, more than one, not united*" is wrong. We therefore look under "* * *Pistil only one, simple or compound.*" The line marked "*a*" is wrong; so, also, is "*b*," but "*c. Herbs: leaves alternate*" describes our plant. "Corolla regular (petals alike) or nearly so" is wrong, so we take the next long line "corolla irregular," etc. Reading under that the five lines beginning with the word "stamens, we have no doubt that the first one, leading to the order Leguminosæ, is the right one. Turning to that order and onward, as before, we find a key in which the following leading lines are correct: "§ 2. Stamens all united or one above distinct: herbs (except some in 3 and 7). * * * * *Leaves pinnate, ending in a bristle, imperfect leaflet or a tendril.* Style flattened, usually twisted half around, one side hairy, 13." Seven pages further on we find: "13. **Lathyrus**, Linnæus." Since this plant is not a native of our country, we do not look further than to note that there are about a dozen kinds which are natives.

When you think you have correctly determined the name of a plant, turn back to the description of the order, and read it carefully once more, so as to be doubly sure. Then if there is any new matter under the order heading read that also. If, for example, you have traced a very common plant to the genus **Phacelia**, § *Euphacelia*, and have concluded that it is the tenth species described on p. 150, turn back to the order **Hydrophyllaceæ**, on page 54. Under the same genus and section there, near the bottom of page 55, you find statements which may change your first decision. What you have taken for *Phacelia tanacetifolia* may be *Phacelia distans* or *Phacelia leptostachya.*

Your labels should be written on slips of paper three inches long, and half as wide. Let the name occupy the upper half, and on the lower half write in small letters where the specimen grew and when you collected it. The label should be fastened in the lower right-hand corner of the sheet on which you mount the plant. Only the left hand end should be pasted down. Paper 17 x 22, cut crosswise, and folded to the size 8½ x 11, is a suitable size for a school herbarium. The plant should be mounted on the third page of the folded half-sheet.

KEY TO WEST COAST BOTANY.

KEY TO THE ORDERS.

.*. Figures in the margin refer to pages. When names are not followed by figures the genus or order indicated is not elsewhere described in this book. Generic names are in italic.

Calyx and corolla together of either more or less than six parts........**CLASS I,** 9
Calyx and corolla together of just six parts: petals never five.
    Stamens six or three.....................................⎫
    Stamens many: sepals three, green............ .... ⎬ **CLASS II,** 16
    Stamens one or two united to the style: ovary inferior...... ⎭
    Stamens many: flowers solitary on long peduncles..........**Papaveraceæ,** 19
    Stamens ten: petal one: a shrub.......................... ......**Leguminosæ,** 30
    Stamens nine, flowers apetalous, small.
        An aromatic tree; flowers greenish...........*Umbellularia.* (*Laurel.*) 71
        Herbs with several or many flowers in involucral cups.....*Eriogonum,* 70
        Herbs with one to three flowers in awned involucral cups. *Chorizanthe,* 70

CLASS I.—EXOGENS OR DICOTYLEDONS.

Calyx and Corolla both present.
    Petals not all united (distinct)..........................**DIVISION 1,** 9
    Petals more or less united (cohering)....................**DIVISION 2,** 13
Calyx and corolla one or both wanting.......... ....**DIVISION 3,** 15

DIVISION 1.—POLYPETALÆ.

A. STAMENS MORE THAN 10.

1. Stamens not adhering to the sepals or petals (ovary not inferior).

          * *Pistils few to many distinct carpels.*

Calyx deciduous, sepals 5: no stipules...................... ....**Ranunculaceæ,** 17

[9]

Calyx persistent, sepals 3 or 4: growing in water.................**Nymphæaceæ,** 18
Calyx persistent, sepals 5 or 10: leaves with stipules....................**Rosaceæ,** 35
Calyx of petal-like sepals: corolla often wanting.................**Ranunculaceæ,** 17

* * *Pistil compound, of 2 or more united carpels as shown by more than one stigma-lobe, stigma, style or cell in the ovary; or by its not being at all one-sided.*

Petals more numerous than the sepals:
  Indefinitely numerous, slender, persistent. Aquatic plants.**Nymphæaceæ,** 18
  Just twice as many (4 or 6): sepals caducous...............**Papaveraceæ,** 19
  Five to sixteen: style 3-8 cleft: fleshy herbs...............**Portulacaceæ,** 26
Petals of the same number (5) as the persistent sepals.
  Leaves opposite: sepals equal...........................**Hypericaceæ,** 25
  Leaves alternate: sepals unequal...........................**Cistaceæ,** 23
  Leaves radical, hollow, 2-appendaged at hooded top.......**Sarraceniaceæ,** 19

2. Stamens and petals on the free or adnate calyx.

Leafless, thorny, fleshy plants: ovary prickly, inferior...............**Cactaceæ,** 43
Leaves mostly opposite, very fleshy: ovary inferior.............. .....**Ficoideæ,** 43
Leaves opposite. Shrub: sepals and petals numerous..........**Calycanthaceæ,** 36
        Shrubs: sepals 4 to 7: flowers white...........**Saxifragaceæ,** 36
Leaves alternate or radical: herbs (ovary not inferior) or shrubs........**Rosaceæ,** 35
Leaves alternate; no stipules: rough herbs: ovary inferior....... .....**Loasaceæ,** 42

3. Stamens on the claws of the petals.

Stamens many, distinct, anthers long: calyx a conical cap: petals 4.**Papaveraceæ,** 19
Stamens many, united into a tube: anthers small: petals 5...........**Malvaceæ,** 27
Stamens 10 to 16, united at base or half way: shrub...............**Styracaceæ,** 50

B. STAMENS 10 OR LESS.

1. Ovary or ovaries superior (i.e., free from the calyx) or mainly so, but sometimes included in the calyx-tube.

* *Pistils more than one, not united.*

Pistils of the same number as petals and sepals.
  Leaves simple, entire, fleshy..........................**Crassulaceæ,** 38
  Leaves pinnate: styles united, globular ovaries distinct.......**Geraniaceæ,** 28

Pistils not of the same number as the sepals and petals.

Two or three.   Shrubs or trees: leaves opposite, compound...**Sapindaceæ,**   30
Herbs; leaves simple.......................**Saxifragaceæ,**   36
Two to ten.   Herbs; leaves pinnate: calyx 10-lobed............**Rosaceæ,**   35
Many.   Stamens on the receptacle.....................**Ranunculaceæ,**   17
Stamens on the calyx: leaves compound, mostly radical..**Rosaceæ,**   35

* * *Pistil only one, simple or compound.*

a.   Shrubs, trees or woody climbers.

Style and stigma one.

Sepals, petals and stamens 6 each, opposite each other......**Berberidaceæ,**   18
Sepals, petals and stamens 4 or 5 each (or stamens 8 in 1st.)
Strongly aromatic or heavy-scented.....................**Rutaceæ,**   29
Not aromatic; leaves simple, opposite................**Celastraceæ,**   29
A vine climbing by tendrils............................**Vitaceæ,**   29
Calyx 2-lipped: petals unequal: stamens 5-8, exserted ....**Sapindaceæ,**   30
Calyx 4-toothed: petals 2: stamens 2 to 4: fruit winged.........**Oleaceæ.**   50
Calyx 4-cleft: petals 4: stamens 6: ovary long-stiped......**Capparidaceæ,**   23
Calyx 4-5 toothed: petals 5: unequal or 1: stamens 10......**Leguminosæ,**   30
Calyx 5-lobed: petals 5, orbicular: stamens 10-15 ....**Rosaceæ,**   35
Sepals 3 or 5, unequal: stamens 4 to 8, united below........**Polygalaceæ,**   24
Styles or Stigmas more than one.
Styles 2 or 3: fruit: 2-winged or inflated: leaves opposite.....**Sapindaceæ,**   30
Styles 3-cleft: stamens 5, opposite small petals..............**Rhamnaceæ,**   29
Stigmas 3: leaves alternate, 3-foliolate or simple..........**Anacardiaceæ,**   30
Stigmas 4 or 5: prostrate stems hardly woody.............**Saxifragaceæ,**   36
Stigma 5-lobed: small shrub: leaves opposite or whorled........**Ericaceæ,**   48

b.   Herbs: leaves mostly or all radical.

Stamens 1 or 3: sepals 2: petals 2 to 5: stigmas 2 or 3......**Portulacaceæ,**   26
Stamens 5, anthers united: lower petal spurred: style 1........**Violaceæ,**   23
Stamens 5, opposite the petals.   Sepals 2: style 3-cleft.....**Portulacaceæ,**   26
Sepals colored, united: styles 5..**Plumbaginaceæ,**   49
Stamens, sepals and petals 5 each: styles 3 or 6: very glandular.**Droseraceæ,**   38
Stamens 5 or 10, on the calyx: styles 2 or 3...............**Saxifragaceæ,**   36
Stamens 8 or 10, on the receptacle: stigma 5-lobed..............**Ericaceæ,**   48
Stamens 10, styles 5: leaves 3 foliolate............**Oxalis in Geraniaceæ,**   28
Stamens 6 united in 3's: sepals 2: petals 4 in unequal pairs..**Fumariaceæ,**   20
Stamens 6: flowers nodding on a scape....**Vancouveria in Berberidaceæ,**   18

POLYPETALÆ.

c. Herbs: leaves alternate.

Corolla regular (petals alike) or nearly so.

Stigma 1, often 2-lobed: stamens 6 (2 and 4) rarely 4............ **Cruciferæ**, 21

stamens 6, equal: ovary on a stipe................. **Capparidaceæ**, 23

stamens 4 to 7 and as many petals on the calyx......... **Lythraceæ**, 38

Stigma 2-lobed: stamens 4: petals 2: sepals 2, white .... **Liliaceæ**. 73

Stigmas 5: sepals 5: petals 5: stamens 10.................. **Geraniaceæ**, 28

Styles 2 or 3: sepals 5: petals 5: stamens 5 or 10: leaves petioled.. } **Saxifragaceæ**, 36

Styles 2 to 5: sepals 5: petals 5: stamens 5: leaves sessile........ **Linaceæ**, 28

Style 2-3 cleft: sepals 2: petals 5 (rarely 2 or 4): fleshy leaves. **Portulacaceæ**, 26

Corolla irregular (petals not all alike): style one.

Stamens 10, included by the cohering lower pair of petals... **Leguminosæ**, 30

Stamens 5, anthers united: lower petal spurred............... **Violaceæ**, 23

Stamens 6, united in 3's: petals 4........................ **Fumariaceæ**, 20

Stamens 6, unequal, distinct or 2 united..................... **Cruciferæ**, 21

Stamens 6 to 8, united: ovary 2-celled: leaves entire........ **Polygalaceæ**, 24

d. Herbs: leaves opposite, simple except in the last.

Style 3-cleft: stamens 3 to 5: leaves a single pair.......... **Portulacaceæ**, 26

Style none, stigmas 3: stamens 10 to 12: petals 6: leaves in 3's. **Papaveraceæ**, 19

Styles 3: flowers sessile; stamens 4 to 7: leaves revolute... **Frankeniaceæ**, 24

Styles 3: flowers in axillary clusters: stamens 3 to 5............ **Ficoideæ**, 43

Styles or stigmas 2 to 5: capsule 1-celled: stamens 10 or 5. **Caryophyllaceæ**, 24

Styles 2: capsule 4-celled: stamens 5......................... **Linaceæ**, 28

Styles 4 or 5: small white flowers in terminal clusters...... **Saxifragaceæ**, 36

Style 1: stamens on the slightly cohering rotate petals...... **Primulaceæ**, 49

Styles and other flower parts each 2 to 5 (stamens rarely twice as many)................................ } **Elatinaceæ**, 26

Styles or stigmas 5: 5 akenes separating when ripe.......... **Geraniaceæ**, 28

2. Ovary and fruit inferior or mainly so.

Shrubs: sepals, petals and stamens each 4 or 5: leaves simple.

Stamens opposite the small clawed petals: style 3-cleft...... **Rhamnaceæ**, 29

Sepals petaloid: ovary globose: styles or stigmas 2........ **Saxifragaceæ**, 36

Sepals, petals and stamens 4 each: the flowers in cymes or in heads with a white involucre....................... } **Cornaceæ**, 44

Herbs. Sepals 5: petals 5: styles 2 to 5: leaves simple......... **Saxifragaceæ**, 36

Flowers or flower clusters axillary

Flower parts in 2's or 4's, small: aquatic: leaves whorled. **Halorageæ**, 38

Flower parts in 4's (rarely in 2's or 6's): style 1........Onagraceæ, 39
Flowers monœcious: climbing by tendrils..........Cucurbitaceæ, 43
Flowers with 2 sepals and 5 petals: fleshy herbs.....Portulacaceæ, 26
Flowers in umbels or heads not axillary
  Flowers in umbels or heads: petals 5: stamens 5.
    Styles 2: fruit dry............................Umbelliferæ, 44
    Styles 2 to 5: fruit juicy........................Araliaceæ, 43
  Flowers in a head with involucre of 4 white leaves.......Cornaceæ, 44

DIVISION 2.—GAMOPETALÆ.

A. **OVARY INFERIOR** (adherent to the calyx) or mainly so.

Stamens 8 or 10: corolla-lobes 4 or 5: shrubs......................Ericaceæ, 48
Stamens 10, those alternate with small corolla-lobes sterile, inflexed......*Samolus*, 142
Stamens 5 (rarely 4) united into a tube.
  Style 2-cleft: flowers in a flower-like head..................Compositæ, 46
  Style and stigma entire: flowers irregular...... ....Lobeliaceæ, 47
Stamens 4 or 5, distinct, growing at the base of the corolla......Campanulaceæ, 48
Stamens on the corolla-tube: leaves opposite or whorled.
  Leaves connate; corolla 4-lobed; stiff, prickly herbs............. ....*Dipsacus*, 46
  Leaves opposite, corolla mostly 5-lobed...................Caprifoliaceæ, 45
  Leaves whorled or sometimes opposite: corolla 4-lobed ....Rubiaceæ, 45
  Leaves unequal: prostrate: calyx corolla-like.....................*Abronia*, 69
Stamens only 3: corolla 5-6 lobed; calyx-lobes minute or none. Herbs.
  Leaves opposite; stamens distinct: erect herbs............Valerianaceæ, 46
  Leaves palmately nerved, alternate: tendril-bearing vines ..Cucurbitaceæ, 43
Stamens apparently 1, really 3 united: flowers monœcious........Cucurbitaceæ, 43

B. **OVARY SUPERIOR** (free from the calyx) or nearly so.

1. **Flowers regular or nearly so.**

        * *Stamens many, united, and adherent to the petals*........Malvaceæ, 27

     ** *Stamens twice as many as the lobes of the corolla.*
Corolla bell-shaped or inflated ovoid................................Ericaceæ, 48
Corolla deeply 5-8 cleft, the base united with the filaments.........Styracaceæ, 50
Corolla 5-cleft: pistils or styles 5; fleshy herbs................ ....Crassulaceæ, 38

* * * *Stamens as many as the corolla-lobes.*

a. *Style 1, stigma 1: leafless, root-parasite*..................................*Pholisma,* 140

b. *Style 1, stigma 1: leaves entire (lobed in the first and last.)*

Leaves mostly radical, reniform: stamens unequal...................*Romanzoffia,* 152

Leaves radical or crowned on rootstocks: flowers salverform... .....**Primulaceæ,** 49

Leaves all radical; flowers spicate, colorless, scarious...... .....**Plantaginaceæ,** 67

        corolla reflexed: anthers purple-black.............*Dodecatheon,* 50

Leaves alternate. Spikes coiled: ovary in 4 parts.............. **Borraginaceæ,** 57

        Flowers rotate to funnelform or tubular.........**Solanaceæ,** 60

        Tall shrub: 3 to 5 calyx-like bracts: flowers yellow..*Fremontia,* 100

        Tall shrub: slightly irregular, nearly white or rose } **Ericaceæ.** 48

        flowers................................. }

        Small herb: minute axillary flowers, the parts in } **Primulaceæ,** 49

        fours................. .... }

Leaves opposite (at least below) entire: juice milky: ovaries 2; stigmas united.

        Flowers white or pinkish in terminal cymose clusters......**Apocynaceæ,** 50

        Flowers in umbels: sepals and petals reflexed or rotate....**Asclepiadaceæ,** 51

Leaves oppposite, ovate to lanceolate, sessile; flowers rotate, axillary.**Primulaceæ,** 49

Leaves clustered at the top of the stem, bracts below: corolla rotate.**Primulaceæ,** 49

c. *Style one or none, stigmas two.*

Leaves opposite or whorled, sessile, entire.......................**Gentianaceæ,** 51

Leaves opposite, lobed: flowers small in spikes...................**Verbenaceæ,** 67

Leaves alternate or radical, 3-foliolate: corolla bearded...............*Menyanthes,* 145

Leaves alternate. Flowers not axillary...... ....**Hydropyllaceæ,** 54

        Flowers in a head with acerose bracts ....*Gilia,* 145

        Flowers funnelform: twining or creeping vines.....*Convolvulus,* 156

Leaves radical: flowers solitary on scapes.........................*Hesperochiron,* 152

d. *Style 1, stigmas 3*...**Polemoniaceæ,** 51

e. *Style 2 cleft*... ....**Hydrophyllaceæ,** 54

f. *Styles 2: leaves simple and alternate or none (i. e. Parasite).*

Flowers solitary, axillary, white: leaves silky...........................*Cressa,* 156

Flowers clustered on filiform, leafless orange or yellow twining stems......*Cuscuta,* 156

Flowers 5 or 6 lines long: shrubs or wood-based herbs........**Hydrophyllaceæ,** 54

g. *Styles 5: calyx not green, petals nearly distinct*........ ......**Plumbaginaceæ,** 49

* * * * *Stamens fewer than the lobes of the regular or slightly irregular corolla.*

Stamens 4: flowers in slender spikes: leaves opposite, lobed........**Verbenaceæ,** 67

Stamens 3: style 3-cleft: sepals 2: leaves opposite, entire ....*Montia,* 97

Stamens 2 or 4: ovary 2-celled........................ ....Scrophulariaceæ, 60
Stamens 2 or 3: scarious corolla, 4-lobed......................Plantaginaceæ, 67

2. Flowers irregular: style 1; stigma entire or 2-lobed.
Leaves or scales not opposite.
    Corolla flattened, heart shaped: stamens 6, united in 3's..........Dicentra, 84
    Corolla curved; leafless root-parasites: stamens 4.......Orobanchaceæ. 65
    Corolla more or less 2-lipped: ovary 2-celled: stamens 2-5.Scrophulariaceæ, 60
    Ovary inferior, stemlike.....................................Lobeliaceæ, 47
Corolla 2-lipped, spurred: ovary 1-celled: stamens 2: aquatic..Lentibulariaceæ, 65
Leaves opposite or whorled: stamens 2 or 4.
    Ovary 2-celled......................................Scrophulariaceæ, 60
    Ovary 4-parted, forming 4 seed-like nutlets....................Labiatæ, 65
    Ovary 2-4 lobed: small flowers in spikes or heads............Verbenaceæ, 67

DIVISION 3.—APETALÆ.

A. OVARY INFERIOR (calyx adherent) or apparently so.

Leaves cordate: calyx 3-lobed: ovary 6-celled..... ....Aristolochiaceæ, 68
Leaves palmately lobed: tendril-bearing vines...................Cucurbitaceæ, 43
Leaves pinnate: calyx-tube 3-4 angled, prickly.................... Rosaceæ, 35
Leaves unequally pinnatifid: calyx-tube in fertile flowers 3-toothed........Datisca, 43
Leaves glaucous: white flowers in clustered umbels..................Comandra.
Leaves small, crenate: capsule axillary, obcordate................Chrysosplenium, 121
Leaves opposite. Calyx salverform: capsule 1-seeded...........Nyctaginaceæ, 68
                 Calyx 4-lobed: stamens 4: flowers axillary....... ....Ludwigia, 125
                 Calyx (corolla) tubular to rotate..............Caprifoliaceæ, 45
Leaves in whorls. Calyx 4-lobed or entire. Aquatic............Halorageæ, 38

B. OVARY SUPERIOR (free from the calyx).

a. Herbs: leaves alternate, radical or in a whorl.

Sepals petaloid, persistent; akene 1, 3-sided or flat...............Polygonaceæ, 69
        persistent: fleshy root parasite, waxy-white bracts....Allotropa, 139
        deciduous: carpels several or many.............Ranunculaceæ, 17
Sepals green: racemes close: capsules flat: 1-2 celled...............Cruciferæ, 21
        minute flowers opposite the leaves....................Alchemilla, 117
Sepals green, 3 larger, spine-tipped; short style, bifid.............Illecebraceæ, 25
Sepals none: dense oblong spike with petaloid involucre..............Anemopsis, 71
        spike naked: radical leaf, 3-foliate. ....Achlys, 82

b. Herbs: leaves opposite, entire: capsule 1-celled except in the last.

Stems square: calyx 4-toothed, with smaller teeth between, axillary ....*Ammania,* 124

Involucre calyx-like or surrounding a head of flowers...........**Nyctaginaceæ,** 68

Flowers small, axillary, sessile, white: stigma 1.........................*Glaux,* 141

Flowers green, terminal: stigmas 3 to 5.....................**Caryophyllaceæ,** 24

Flowers purplish, in axillary clusters, minute....................**Illecebraceæ,** 25

Flowers axillary: capsule 3-5 celled................. ....**Ficoideæ,** 43

c. Shrubs or trees: leaves alternate, entire (except in the last): flowers perfect.

Calyx tubular, bearing the stamens: akene plumose-tailed ....*Cerocarpus,* 115

Calyx 6-parted, yellowish: leaves very aromatic..... .... **Lauraceæ,** 71

Calyx 4-5 cleft, greenish: fruit berry-like, 2-4 seeded..............**Rhamnaceæ,** 29

Calyx 3-4 lobed, yellow: stamens 6 to 12, exserted.................*Dirca palustris.*

Calyx 5-cleft, large, yellow: stamens 5, united................... ..... **Sterculiaceæ,** 28

d. Trees or woody climbers with opposite pinnate leaves.

Trees: flowers diœcious, winged fruit in drooping panicles..............**Oleaceæ,** 50

       flowers perfect: fruit 2-winged, 2-seeded...................**Sapindaceæ,** 30

Climbers: sepals 4: stamens and pistils many, akenes tailed.............*Clematis,* 79

e. Diœcious shrubs with drooping silky grey aments.........................Garrya, 131

CLASS II—ENDOGENS OR MONOCOTYLEDONS.

A. OVARY SUPERIOR.

Pistils 8 to many distinct or united carpels; flowers in whorls........**Alismaceæ,** 72

Pistil 3-carpeled, ovary 3-celled, or at least 3-sided....................**Liliaceæ,** 73

Pistil 2-celled: red flowers in an umbel....................... ....*Clintonia,* 188

     small perianth, 4-parted: stamens 4.....................*Maianthemum,* 184

B. OVARY INFERIOR.

Flowers irregular: anthers 1 or 2 on the pistil....................**Orchidaceæ,** 72

Flowers regular: anthers 3, extrorse.......... ....**Iridaceæ,** 73

DESCRIPTIONS OF THE ORDERS

OF

WEST COAST PLANTS

AND

SPECIES NAMED SINCE 1886.

DIVISION I. POLYPETALÆ.

RANUNCULACEÆ.

Herbs or shrubs, with colorless juice: foliage various: stipules none: organs of the flower free and distinct; sepals, petals and carpels, few or many, not united: stamens numerous: petals sometimes wanting, then the sepals are usually petaloid; anthers short and adnate; seeds with minute embryos in fleshy albumen. Key to genera and species, p. 79.

Anemone nemorosa of the Atlantic slope differs from the Pacific form, which is considered worthy to be called a separate species by some botanists, viz: **A. Grayi.**

Ranunculus maximus, Greene, has akenes, with nearly straight long beaks like those of **R. orthorhyncus**, of which species it may be a variety.

Thalictrum hesperium, Greene, is distinguished from **T. polycarpum** by an ill-scented, not aromatic, odor.

T. Cæsium, Greene, of the Sierra Nevada foot-hills, is odorless.

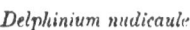

Delphinium nudicaule.

Botany —2

DELPHINIUM.

D. hesperium, Gray. · Stem slender: herbage canescent: leaves small, much dissected into linear obtuse lobes: racemes long, dense: spur of the deep blue or pinkish flowers, stout, straight, about as long as the sepals: carpels hairy. Flowering in June.

D. Hanseni. Greene. Distinguished by white burr-like seeds.

D. Blochmanæ, Greene. A foot high: large flowers in a strict short raceme, the inner light blue sepals, and the white petals with conspicuously crisped margins. San Luis Obispo Co.

D. distichum, Geyer. Flowers somewhat 2-ranked: spur 9 lines long, straight, twice the length of the sepals. Northern Coast.

D. ulignosum, Curran. Leaves fan-shaped, 3-cleft, the segments 3-toothed: the rather large deep blue flowers with straight spurs equaling the sepals. Wet ground, Lake Co.

COPTIS.

C. laciniata, Gray. Leaves trifoliolate, ovate, nearly 3-parted, the segments incised: sepals and petals linear-attenuate. Rare from Humboldt Bay northward.

BERBERIDACEÆ.

Shrubs or herbs, with compound alternate exstipulate leaves; flowers remarkable for having the bracts, sepals, petals and stamens before each other, instead of alternating. Key to genera and species, p. 82.

Our species of Berberis are very different in appearance from the scarlet-fruited Barberry of Europe and the Atlantic States, which is often cultivated.

Vancouveria parviflora, Greene, may be considered a variety of **V. hexandra.**

· NYMPHÆACEÆ.

Aquatic perennial herbs, with peltate or deeply cordate leaves; solitary axillary perfect flowers on long peduncles. Stamens numerous.

Water-Shield (*Brasenia peltata,* Pursh.) May be found in ponds. Its elliptical, peltate, floating leaves (green above and brownish-red beneath) and its jelly-coated stems characterize it quite well enough.

The *Yellow Pond Lily* (*Nuphar polysepalum*) Engl.) is more common. It is **Nymphæa polysepala,** of the *Bay-Reg. Bot.* Key to genera and species, p. 82.

SARACENIACEÆ.

A small order of bog plants, remarkable for their pitcher-shaped, tubular and hooded leaves which entrap insects. Key to genera and species, p. 83.

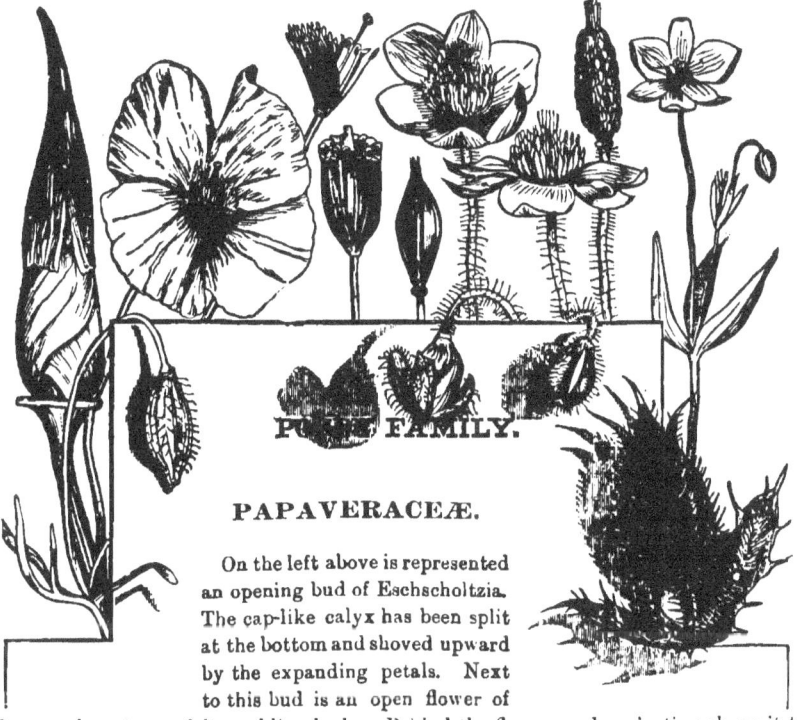

PAPAVERACEÆ.

On the left above is represented an opening bud of Eschscholtzia. The cap-like calyx has been split at the bottom and shoved upward by the expanding petals. Next to this bud is an open flower of Meconopsis and one of its nodding buds. Behind the flower, and projecting above it to the right, is a stem from which the petals have just fallen. The slender filaments bending to one side, as they often do, show the curious pistil, which in time becomes the pretty fluted capsule seen below. To the right of the Meconopsis pod is the three sided capsule of Platystigma lineare. The stem should have a few hairs upon it. The two flowers with hairy stems, the nodding buds below, and the rough seed pod above, belong to Platystemon. Observe the three caducous sepals, just ready to drop from the opening bud. The smooth plant on the right is Platystigma Californicum. If you choose you may call this the Smooth Platystigma, and the other species, with the triangular pod, Hairy Platystigma. The exceedingly prickly Bristly Argemone is represented on the right, below, by a bud and a couple of bracts. A pistil with its white prickles is imperfectly shown against one of the bracts.

The Order **Papaveraceæ** is characterized by flowers with 2 or 3 caducous sepals, twice as many free petals in two sets, indefinite, usually numerous, free stamens, and a compound pistil. In *Eschscholtzia* the sepals are united, and the stamens adhere to the claws of the petals.

This small but interesting order of plants, with the exception of one species, is confined to the northern hemisphere. Fifteen species, belonging to eleven genera, are natives of the United States, and several European species have become naturalized. Eschscholtzia and Platystemon are the most widely distributed of the California genera. Key to genera and species, p. 83.

PAPAVER and MECONOPSIS.

Our species of **Meconopsis** is put by Greene with the true poppies in the genus **Papaver.** He thinks with Bentham that the small erect flowering form is a separate species, viz: **P. crassifolium.**

Two more species of **Papaver** may be added.

P. Californicum, Gray. 1–2½ feet high; leafy below: peduncles long: corolla 2 inches broad, brick red with green spots at the base: stigmas sessile and radiating.

P. Lemmoni, Greene. Larger: stigmas 7–10, their lower half sessile and radiant upon the pod, the upper half coherent and forming a conical apiculation.

PLATYSTEMON, PLATYSTIGMA and ESCHSCHOLTZIA.

Prof. Greene unites the genus **Platystigma** with **Platystemon** and adds a new species.

P. Torreyi, Greene. Glabrous: the three carpels united to form a slender, twisted pod.

Botanists will probably never agree about the number of species of **Eschscholtzia.**

FUMARIACEÆ.

Tender herbs with dissected compound leaves, and irregular hypogynous flowers, the parts in twos, except the 6 diadelphous stamens.

Key to genera and species, p. 84.

Dicentra is **Capnorchis** in Greene's *Bay Region Botany.*

CRUCIFERÆ.

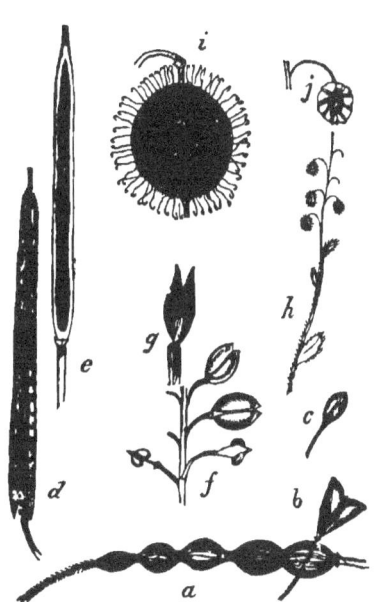

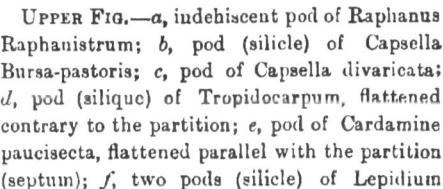

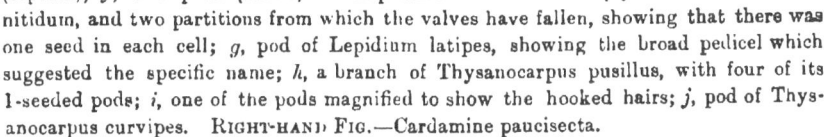

UPPER FIG.—*a,* indehiscent pod of Raphanus Raphanistrum; *b,* pod (silicle) of Capsella Bursa-pastoris; *c,* pod of Capsella divaricata; *d,* pod (silique) of Tropidocarpum, flattened contrary to the partition; *e,* pod of Cardamine paucisecta, flattened parallel with the partition (septum); *f,* two pods (silicle) of Lepidium nitidum, and two partitions from which the valves have fallen, showing that there was one seed in each cell; *g,* pod of Lepidium latipes, showing the broad pedicel which suggested the specific name; *h,* a branch of Thysanocarpus pusillus, with four of its 1-seeded pods; *i,* one of the pods magnified to show the hooked hairs; *j,* pod of Thysanocarpus curvipes. RIGHT-HAND FIG.—Cardamine paucisecta.

Herbs with pungent watery juice. Sepals 4. Petals 4, with blade narrowed into a claw, the lamina spreading to form a cross, sometimes unequal, rarely wanting. Stamens 6, two of them inserted lower down on the receptacle, and usually shorter than the other four, rarely only 4 or 2. Ovary 2-celled by a thin partition, rarely 1-celled. Leaves alternate, and flowers usually in racemes without bracts.

A careful examination of the fruit is usually necessary for a determination of the species in this difficult order. Key to genera and species, p. 84.

In **Tropidocarpum** only are the flowers solitary and axillary.

T. capparideum, Greene, is distinguished by inflated pods much shorter than represented at *d* in the figure, and opening from above. Perhaps only a variety.

Under **Thelypodium** the following species may be added:—**T. Hookeri,** Greene. Distinguished from *T. flavescens* by broader sepals which with the pedicels are smooth; broader claw and narrower blade of petals and longer pods: perhaps only the Monte Diablo form of the latter species. **T. rigidum,** Greene. Stout and very rigid, 1–3 feet high; pods nearly sessile, 1¼ inch long, rigid, sharply tipped with the short style. Monte Diablo.

The perennial species of Cardamine in *Bay Region Botany* may be considered forms of *C. paucisecta*. The species of Streptanthus are so variable that botanists differ widely as to their limits. Their irregular flowers make them worthy of the special attention of students who are interested in the relations between flowers and insects. What insects are accommodated by this irregularity? The flowers vary from white to nearly black in color. Are these forms—for they certainly were one form and color not long ago—found growing together? Are they visited by different insects?

Prof. Greene adds in *Bay-Reg. Bot.* seven species, viz:

S. barbiger. Stem leaves linear, entire; sepals equal, greenish, the acuminate tips whitish and recurved: petals white: filaments dark purple. Near St. Helena.

S. suffrutescens. Perennial stems with a stout leafy persistent trunk, bearing longer flowering branches: stem-leaves cuneate-obovate, coarsely serrate; floral leaves, round cordate or narrower. Collected by Bioletti on Hood's Peak, Sonoma Co.

S. albidus. Distinguished from **S. niger** by white not dark purple sepals, the anthers of the united pair of stamens bearing pollen. Found only on Oak Hills near San Jose.

S. Mildredæ. Slender, less than two feet high: small flowers, very dark metallic-purple: slender petals, white-margined. Mt. Hamilton.

S. Biolettii. Similar to S. glandulosus: more slender racemes, not one-sided: sepals darker, smaller. Hood's Peak.

S. pulchellus. A foot or less in height, much branched: slender leaves with a few coarse teeth: calyx lilac purple, the sepals nearly equal. Mt. Tamalpais, on south side.

S. secundus. Lower leaves pinnately lobed or toothed: racemes of flesh-colored flowers, one-sided: lower sepal clawed. North side of Mt. Tamalpais.

These, excepting possibly the first two, may be considered as forms of **S. Glandulosus.**

Common water-cress, naturalized everywhere, is **Nasturtium officinale.** The yellow-flowered species of that genus are put in Gesner's genus, **Roripa,** by Greene in his *Bay-Reg. Bot.,* and one species is added, viz:

R. dictyota. Two to four feet high: pods ovate lanceolate. Marshes of the lower Sacramento.

R. lyrata may be a variety of *N. curvisilequa.*

High authority enables us to shorten the cumbrous name, *Capsella Bursa-Pastoris* into *Bursa pastoris,* which exactly means shepherd's purse.

Cheiranthus asper, C. & S., is **Erysimum asperum** of *Bay-Reg. Bot.,* but there confined to the tall form with usually orange flowers. According to Greene (Pittonia, Vol. III, Part 15, p, 131) the low perennial common along the coast with cream-colored to sulphur-yellow flowers is **C. capitatus,** Douglas.

C. angustatus, Greene. Slender perennial 2 feet high, or more; leaves narrow entire or nearly so. Corolla large, yellow, a little one-sided. San Joaquin River.

C. Calfornicus, Greene. Stout, biennial, strongly angled; leaves runcinately toothed; pods sharply 4-angled.

C. Occidentalis, Watson. A low annual with winged seeds. Or. and Washington.

Our species of **VESICARIA** may now bear the generic name **LESQUERELLA.**

Caulanthus Lemmoni, Wat., Glaucous, 1 or 2 feet high: sepals brownish purple, spreading or reflexed; narrow petals undulate 6 to 8 lines long, white, brown veined pods 2½ to 5 inches long: stigmas spreading.

CAPPARIDACEÆ.

Herbs or shrubs with alternate leaves: differing from Cruciferæ in equal stamens, pods on stipes and pedicils commonly bracteate. Plants of this order are mostly in the warm-temperate and tropical regions, while the Cruciferæ are mostly found in the cool, temperate and polar regions. Key to genera and species, p. 90.

CISTACEÆ.

Flowers perfect and regular. Sepals 5, persistent; and two of them smaller, wholly exterior, and bract-like. Petals 5, usually ephemeral. Stamens indefinite, with filiform filaments; anthers short. Style one. Capsule 3-valved. See p. 91.

VIOLACEÆ.

Herbs distinguished by the irregular one-spurred corolla of five petals, 5 stamens, adnate introrse anthers conuiving over the pistil, which has a club-shaped style with a

24 CARYOPHYLLACEÆ.

one-sided stigma, a one-celled ovary, forming a capsule, which splits at maturity into three parts. Key to genera and species, p. 91.

Add after the third species of Viola **V. Howellii**, Gray. A blue violet with a very short, thick spur. Oregon.

POLYGALACEÆ.

Herbs or shrubs, with simple entire exstipulate leaves, remarkable for the flowers which appear like those of the Pea family, but the structure is very different.—Leaves simple, entire: stamens less than ten: pistil 2-carpelled. Key to genera and species, p. 91.

FRANKENIACEÆ.

Sessile, small, opposite leaves: small flowers, sessile, in the axils of the numerous branches: ribbed calyx, tubular. See p. 92.

CARYOPHYLLACEÆ.

Herbs with regular and mostly perfect flowers, persistent calyx, its parts and the petals 4 or 5 and imbricated, or the latter sometimes convolute in the bud, the distinct stamens commonly twice as many as the petals, ovary 1-celled with a free central placenta. Stems usually swollen at the nodes. Leaves opposite, often united at the base by a transverse line, in one group, with interposed scarious stipules. Styles 2 to 5, mostly distinct. Fruit a capsule opening by valves, or by teeth at the summit. Flowers terminal, or in the forks, or in cymes. Key to genera and species, p. 92.

SILENE.

Silene racemosa, Otto, an annual, has deeply bifid white petals: fragrant. Introduced. Berkeley.

S. multinervis, Watson. Ovoid calyx conspicuously 20–25 nerved: small purplish petals, not appendaged.

S. inflata, Smith. Slender perennial: calyx ovoid: large white petals, bifid. Naturalized. Vallejo.

S. Luisana, Wat. Perennial, glandular-pubescent: calyx teeth, with membranous ciliate margin: white petals, bifid. San Luis Obispo, Monterey.

S. Bernardina, Wat. Perennial, glandular-pubescent: petals greenish, cleft into 4 equal narrow lodes, appendages nearly half the length of the blade, 2-parted, the inner segment lacerate. Long Meadow, Tulare Co.

S. Suksdorfii, Robinson. Low, densely matted, alpine, stem leaves usually 2 pairs, linear, 3-7 lines long, a line wide: radical leaves, crowded: petals white, slightly bifid.

POLYCARPON.

Polycarpon tetraphyllum, Linn. Lower leaves nearly in whorls of 4: branches 3-7 inches long. Naturalized.

LYCHNIS.

Lychnis Githago, Lam. (*Agrostemma Bay-Reg. Bot.*) Corn-Cockle. Erect 2-4 feet high: slender leaves, united at base: calyx over an inch long, the slender lobes surpassing the purple unappendaged petals. A naturalized weed.

Saponaria Vaccaria is **Vaccaria vulgaris**; **Stellaria** is **Alsine**; **Sagina** is **Alsinella**; and **Lepigonum** is **Tissa** in *Bay-Reg. Bot.* Prof. Greene describes nine species of the last growing in Bay region. All these, with **L. gracile**, may be considered varieties of two species, or even of one. Authorities have not decided whether **Tissa** or **Buda** shall take the place of **Lepigonum** as the generic name of these interesting plants.

ILLECEBRACEÆ.

Leaves opposite or alternate, pungent, small: flowers inconspicuous, axillary; petals wanting or rudimentary: style undivided or 2-cleft: fruit 1-seeded. Key to genera and species, p. 96.

Add: **Paronychia Chilensis**, D. C. Leaves opposite, on tough, short-jointed, suffrutescent stems: purplish calyx, minute. Presidio, San Francisco.

Herniaria cineria, D. C. Slender annual: ashy leaves and flowers, minute. Introduced. Monte Diablo.

HYPERICACEÆ.

Herbs or shrubs, with opposite, entire, punctate leaves, no stipules and perfect flowers with 4 or 5 petals and numerous stamens, the fruit a septicidal many-seeded capsule. Calyx of 4 or 5 persistent sepals. Filaments mostly in 3 sets. Styles 2 to 5, usually distinct. Key to genera and species, p. 98.

HYPERICUM.

Hypericum mutilum, Linn., has been found in Solano County by Jepson. Stamens only 5-12, distinct.

PORTULACACEÆ.

Succulent herbs, with simple and entire leaves, and regular but unsymmetrical perfect flowers; the sepals only 2, the petals 2 to 5 or more; the stamens opposite the petals when of the same number: the ovary 1-celled. Stamens sometimes indefinitely numerous, commonly adhering to the base of the petals, these sometimes united at the base. Style 2–8 cleft. Stipules none. Key to genera and species, p. 96.

CALANDRINIA.

Calandrinia Howellii, Wat., is distinguished from *C. cotyledon* by narrower leaves, with scarious margin crisped.

Calandrinia elegans, Spach., is the larger, stouter form of *C. Menziesii*, Hooker. Greene confines the latter to the very small-flowered dwarf form.

CLAYTONIA.

Claytonia spathulata is equivalent to C. gypsophiloides of *Bny-Reg. Bot.* It has rose-purple petals, three times as long as the calyx. C. exigua is probably C. spathulata of Prof. Greene's book. He adds C. nubigena, which is described as similar to C. perfoliata, but smaller, with linear radical leaves.

a

b

C. Hallii, Gray, comes next to *C. Chamissonis.* Leaves, 2 or 3 pairs: seeds 1 or 2. For C. Nevadensis substitute C. asarifolia, Bongard.

Perhaps Claytonia and Montia will be united under the latter (older) name.

Spraguea, too, for a similar reason may be merged in Calyptrideum. C. tetrapetatum of *Bny-Reg. Bot.* is C. quadripetalum of this book.

Montia parvifolia, Greene, has petioled leaves and larger flowers than M. fontana.

ELATINACEÆ.

Low annuals growing in water or wet ground. Flowers minute, axillary. Key to genera and species, p. 98.

MALVACEÆ.

Herbs or shrubs with alternate stipulate leaves; distinguished by the valvate calyx, convolute petals, their bases or short claws united with the base of a column of many united stamens, these with reniform anthers. Calyx 5-cleft or parted, persistent, with sometimes a calyx-like involucel of bracts. Petals 4, usually withering without falling off. Pistil usually either a ring of ovaries around a projecting receptacle or a 3–10 celled ovary: styles united at least at the base. Leaves usually palmately ribbed. Flowers axillary.

Cotton is the most notable plant in this order. Key to genera and species, p. 98.

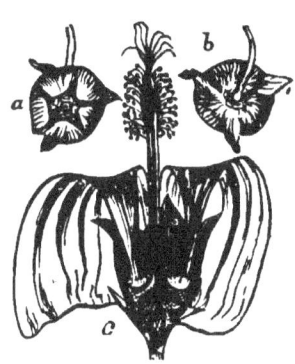

LOWER FIG.—*a.* Fruit of Malva borealis. *b.* Same, showing the bracts of the persistent calyx. *c.* Kellogg's Lavatera. (L. assurgentiflora.)

MALVA.

Malva parviflora, Linn., is distinguished from **M· borealis** by the calyx lobes spreading away from the fruit and the smaller flowers.

Malva rotundifolia, Linn., is distinguished from these two species by akenes rounded on the back, so as to make the fruit somewhat scalloped.

The so-called *Marsh mallow* is not a *Malva*, and would better be called *Marsh Hollyhock.* It probably does not grow on this coast.

SIDALCEA.

S. Hendersoni, Wat. Simple stems, 3–4 feet high, glabrous: flowers 9-12 lines long: carpels smooth and beaked. Oregon.

S. Hickmanii, Greene. Rough with stellate hairs: stem leaves round fan-shaped: racemes numerous, axillary and terminal, few flowered: short pedicels, subtended by 3 slender bractlets, 5 lines long: corolla an inch broad: akenes nearly orbicular. Canyons of Salinas Valley.

S. secundiflora, Greene, is var. *minor* of *S. diploscypha*, in this book.

MALVASTRUM.

M. Parryi, Greene. Annual: purplish and rough hairy branches, 1-2 feet long: hoary with stellate hairs: flowers mostly solitary on peduncles 1-4 inches long: involucel of 3 slender bracts: petals deep purple, 5-9 lines long: carpels 15-20.

M. arcuatum, Greene, similar to M. Thurberi, but the leaves with rounded, not angular lobes, and only half as large.

Malvastrum is put under **Malveopsis** in *Bay-Reg. Bot.* **Malveopsis fascicula-tum** of that book is **Malvastrum Thurberi** of this.

Abutilon Theophrasti, Medic, a large weed, native of India, is reported as introduced about Santa Rosa. It has large velvety leaves and yellow flowers.

MODIOLA.

M. multifida, Moench. Calyx bracts 3: leaves deeply cleft, the lobes toothed: corolla a half inch broad, rose color. Streets of San Jose. *Chas. Jared.*

STERCULIACEÆ.

Fremontia grows on Mt. St. Helena, near the Great Western mine. Specimens and description furnished by Miss L. A. Martin show the form there to be smaller and much less beautiful than the Sierra Nevada form. See key to genera and species, p. 100.

LINACEÆ.

Parts of the flower 5, except sometimes in the pistil. Filaments united at the base, with commonly alternating teeth. Styles 5, or sometimes only 2 or 3, distinct or united. Stigmas capitate or oblong: ovary globose. Seeds twice as many as the styles. Herbs with sessile entire leaves without stipules, and cymose or panicled flowers. Key to genera and species, p. 100.

Linum perenne is **L. Lewisii,** in *Bay-Reg. Bot.*

GERANIACEÆ.

Flowers perfect on axillary peduncles, regular (in our species) and symmetrical, the parts in fives. Stamens mostly in two sets, those alternate with the petals sometimes sterile. Ovary deeply lobed, with a prolonged axis, or 5-celled. Key to genera and species, p. 101.

According to Greene in *Bay-Reg. Bot.* the following geraniums of the Old World have become established.

G. dissectum, Linn. Over a foot high, distinguished from **G. Carolinianum** by larger red-purple flowers, the petals more deeply emarginate.

G. molle, Linn. Soft hairy, slender: flowers small, rose-red.

G. retrorsum, L'Her. Perennial, very small flowers.

Greene makes a new native species of *Erodium* (**E. Californicum**) distinguished from **E. macrophyllum** by deep rose-red flowers, instead of dull-white.
Limnanthes is **Flœrkea** in *Bay-Reg. Bot.*, and **Oxalis** is **Oxys**.
Oxalis Oregana has usually solitary flowers on 2-bracteate scapes. **O. trillifolia** has flowers in bracteate umbels.

RUTACEÆ.

Pellucid or glandular-dotted aromatic leaves, along with definite hypogynous stamens characterize this order, although some of the orange-tribe have many stamens. Shrubs or trees. Key to genera and species, p. 101.

CELASTRACEÆ.

Shrubs with simple undivided leaves and dull-colored or white flowers. Sepals, petals and stamens 4 or 5: stamens on a disk. Key to genera and species, p. 102.

RHAMNACEÆ.

Shrubs or small trees, with simple undivided leaves, small and often caducous stipules, and small regular flowers, the stamens borne on the calyx and alternate with its lobes: ovary 2 to 4-celled. Flowers often apetalous: a conspicuous disk adnate to the short tube of the calyx: petals often clawed: style or stigma 2–4 lobed: fruit berry-like or dry, containing 2 to 4 seed-like nutlets. Key to genera and species, p. 102.

According to Greene, **Rhamnus ilicifolia**, Kellogg, may be distinguished from **R. crocea** by its greater size, the latter only 2 or 3 feet high, and the parts of the flower in 5's, not 4's.

Rhamnus Purshiana is *Cascara sagrada*, by far the most notable medicinal plant of this coast.

R. Californica is absurdly called *California Coffee*.

VITACEÆ.

This small order is represented on this coast by one species, **Vitis Californica**, known as the *California grape*. In the Atlantic states there are half a dozen species. In Europe probably no native species. Virginia Creeper, *Ampelopsis quinquefolia*, commonly cultivated, belongs to this order. Botanists think all the wine and raisin grapes cultivated in Europe are varieties of one or two species. *Isabella, Catawba* and other

30 LEGUMINOSÆ.

cultivated grapes, known as American grapes, are varieties of *V. labrusca* of the Atlantic States.

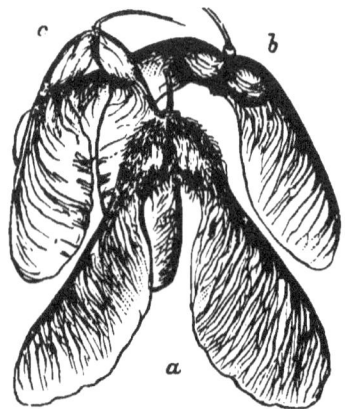

In the figure *a* represents the fruit of *Ac r macrophyllum*, *b* the wider spreading samara of *Acer cercinatum*, and *c* the closer wings of *Negundo Californicum*. The first has hairy carpels; the second is smooth, and the last slightly hairy.

SAPINDACEÆ.

Trees or shrubs with opposite, compound or palmately lobed leaves: sepals 5: petals 4 or 5: pistil 2–3 carpelled.

The flowers and fruit of our common Buckeye are very interesting. A bunch of several hundred flowers usually produces but one fruit; and this, formed of three carpels containing six ovules, rarely ripens more than one seed. The genus **Negundo** is united with **Acer** in *Bay-Reg. Bot.*, and **Æsculus** is **Hippocastanum**. Key to genera and species, p. 103.

ANARCARDIACEÆ.

Shrubs or trees with resinous and often poisonous juice, alternate leaves and small flowers. **Rhus diversiloba** (*Poison Oak*) is the most common species. A poisonous species in Japan furnishes the remarkable Japan varnish. Key to genera and species, p. 104.

LEGUMINOSÆ.

The single and simple free pistil becoming a legume in fruit, the alternate leaves with stipules, and in our genera, the papilionaceous corolla with 10 stamens, mark this order, one of the largest and most important in the vegetable kingdom.

Flowers irregular. Calyx 3-4 cleft or toothed, persistent. Corolla of 5 petals, the upper large and always external, covering the lateral pair in the bud, and these covering the lower pair which are more or less united, forming a keel which incloses the stamens and pistil. Filaments 10, rarely 5, commonly united around the pistil, either all united or nine and the upper one free. Ovary forming a pod with a single row of seeds attached to one side: style usually inflexed or curved. In *Cercis* the upper petal is small and enclosed by the wings. In *Amorpha* there is but one petal.

Suborder **Cæsalpineæ** is marked by the upper petal enclosed and distinct stamens.

Suborder **Mimoseæ** has regular flowers and usually many conspicuous stamens.

Fɪc. A. On the left is *Hosackia subpinna'a*, showing a full-grown pod and a flower as seen from above. On the right is a pod and flowers of *Hosackia Purshiana*. At *a* is a single flower with its bract as seen from the front. The lower leaves and bracts are larger.

Fɪɢ. B. A head of *Trifolium fucatum*, with all but three of the flowers removed, showing the common receptacle and the involucre.

Fɪɢ. C. An axillary spike of *Astragalus didymocarpus*, with ripe fruit. Below is one of the pods magnified.

This order is remarkable for the number of useful and beautiful plants which belong to it. Pease, beans, lentils, peanuts, clover, alfalfa, etc., furnish food for man and domestic animals. Tropical plants of this order supply, among others, the following articles of commerce: Gum arabic, gum senegal, gum copal, dragon's blood, indigo, logwood, brazilwood, rosewood, tamarind. Many species have medical value, as senna, catchu, copaiba, etc.

There are over 6,000 species of leguminous plants, mostly tropical. About 350 species are natives of the United States, more than half of which are found in California. Only 4 or 5 species are common to this coast and the Atlantic States, and these have forms peculiar to each coast. Our 180 species are grouped under 14 genera, while the 150 species of the East (*i. e.*, the Mississippi States and eastward to the Atlantic), represent 50 genera. There are about 40 species of lupine, and the same number belonging to the genus Astragalus, growing within the limits of this State. Only 2 kinds of the former and 4 of the latter grow east of the Mississippi. The latter is the largest American genus of the order, the species within the United States numbering about 150, nearly all of which belong west of the Rocky Mountains. We.have about 25 kinds of clover; only 3 or 4 species are natives of the East. Hosackia, numbering 28 species in our whole country, 25 of which grow here, is not represented in the East at all. On the other hand, the large genus Desmodium, numbering in the East 19 species, has no representative west of the Rocky mountains. Pickeringia is probably not found beyond the boundary of California. The great Australian genus Acacia, numbering there nearly 300 species, is represented in Southern California by a small tree (*A. Greggii*), and in the East by an herb. Possibly 30 species are cultivated for shade trees. Honey Mesquit, or Algaroba (*Prosopis juliflora*) and Screw-pod Mesquit, or Tornilla (*P. pubescens*), are small trees of Southern California. Prosopis and Acacia belong to the suborder Mimosæ. All the plants here described (except *Cercis*) belong to the suborder Papilionaceæ, which is distinguished by flowers, like those of the pea, as before described.

The devices for securing cross-fertilization in this order are often very remarkable. Key to genera and species, p. 104.

LUPINUS.

A.   Perennials, more or less shrubby.

L. jucundus, Greene.   Long peduncled racemes: keel naked, banner notably smaller than the other petals.

L. eminens, Greene.   Almost arborescent: short and dense racemes, short peduncled: pods villous, nearly erect, about 4-seeded.

L. variicolor, Steud.   Woody basal branches short, slender: raceme short: flowers large: keel ciliate, pods large.

The above, with **L. albifrons,** Benth., may be considered as varieties of **L. Chamissonsis,** of which Greene says, "Apparently confined to the sand dunes of the San Francisco peninsula and Pt. Reyes.

L. propinquus, Greene.   Near **L. arboreus,** but small: racemes short: petals violet.   A seaboard species.

B.   Perennial herbs: flowers large.

L. formosus, Greene.   Leaflets 7-9, linear-lanceolate, very acute: flowers rich violet: keel naked.   Common in fields.

L. sericatus, Kellogg,   A foot or less in height, silky-canescent, leafy; leaflets 7, spatulate oblong: keel slender, pointed, lightly ciliolate.   In the Mt. Helena region:

L. latifolius, Agardh.   Erect, 2–4 feet high, the stem dark green and shining, leaflets 5–7, broadly oblanceolate, thin, ciliolate: blue flowers changing to dull brown. Has been considered a form of **L. rivularis.**

E. Annuals: flowers verticillate, small, deep-blue.

L. polycarpus, Greene. Distinguished from *L. micranthus* by stoute.·, rather suc-culent stem: upper calyx-lip with teeth not divergent, lower slightly notched: corolla smaller: base of the keel exposed: ridged pods, slightly falcate, 7-9 seeded.

L. pachylobus, Greene. Stout, ridged, barely a foot high, hairy: petioles long and slender: leaflets 5-7, 9 lines long: whorls 2-4 on stout peduncles: very hairy pod, 4-6 seeded.

L. carnosolus, Greene. Stout and succulent stem, usually simple: raceme loose: keel hairy in the middle.

TRIFOLIUM.

Greene makes five species of *T. Macræi* and its varieties, viz:

T. Macræi, H. & A. Much branched, decumbent or almost prostrate, more or less hairy: leaflets cuneate-oblong, denticulate above the middle: heads nearly or quite sessile, 6 lines, or less, long: calyx teeth densely plumose, hairy, nearly equaling the purplish corolla: pod 1-seeded. A variety with smaller heads on very long peduncles. Variable.

T. dichotomum, H. & A. Stouter, often erect, flexuous and dichotomous: heads long peduncled: calyx teeth equaling the red-purple corolla.

T. amœnum, Greene. Erect, taller: heads 18 lines long: plumose calyx teeth, much shorter than the conspicuous light rose-purple corolla.

T. columbinum, Greene. Erect, rather silky: silky plumose dove-colored calyx teeth, concealing the minute corolla.

T. olivaceum, Greene. Olive-green heads, in which the corollas are concealed by the calyx teeth, which are nearly naked at the setaceous tips.

The clovers with flat laciniately-cleft involucres, are put under the following names in *Bay-Reg. Bot.:*

T. Wormskjoldii, Lehm. Perennial: leaflets obovate-oblong, denticulate: heads hemispherical, an inch or more broad, the involucre half as broad: 10-striate calyx tube. The teeth much longer and sometimes cut into setaceous divisions. [This is *T. involucratum,* var. **heterodon,** of this book.]

T. variegatum, Nutt. Annual: leaflets obcordate to obovate-oblong: calyx-tube about 15-nerved, the teeth broader than in the last. [A form of **T. involucratum,** but the var. **melanthum** seems to be **T. tridentatum,** var. **melanthum,** and var. **major** is var. **obtusiflorum.**]

T. appendiculatum, Loja. Flaccid, diffuse: keel of the corolla rostrate-attenuate.

T. oliganthum, Steud., and the varieties, are in this book under *T. pauciflorum.*

T. roscidum, Greene. Erect purple stems, flexuose: leaves soft, hairy, and very clammy. [This, with **T. obtusiflorum,** have heretofore been by good authorities considered as varieties or forms of **T. tridentatum.**]

Botany — 3

34

T.o the cup clovers, with concave involucres and usually inflated, corollas in fruit, Prof. Greene has added these species.

T. Grayi, Loja, which is **T.** barbigerum, var. **Andrewsii**, Gr. Seemingly a good species.

[**T. flavulum**, Greene; **T. virescens**, Greene, and **T. Gambelii**, Nutt., are forms of **T. fucatum**, Lindl., which varies from wet meadow-land forms, a foot or more high, with heads of 20 flowers, 2 inches broad, to the upland form **T. Gambelii**, Nutt., scarcely half as high, bearing heads with only 3 or 4 flowers.]

T. hydrophyllum, Greene. Flaccid branches 1–2 feet long: stipules ovate, entire: leaflets linear to oblong, those of the lower leaves narrowest: peduncles short: involucre of about 5 bracts: calyx-teeth very long. Apparently a good species hitherto called **T. amplectens** or a variety of **T. depauperatum**, to which the following may be referred.

T. laciniatum, Greene. Upper leaves broad truncate, 3 dentate at apex, laciniately toothed or pinnatifid.

HOSACKIA.

In *Bay-Reg. Bot.* **Lotus Americanus** is **H. Purshiana** of this book; **L. Wrangelianus** is **H. subpinnata**; **L. humistratus**, Greene, is **H. brachycarpa**; **L. denticulatus**, Greene, is **H. subpinnata**, var. **major**, which seems to be a good species; **L. micranthus** is **H. parviflora**; **L. rubellus** and **L. nudiflorus** are **H. strigosa**—the former a coast form, the latter of the interior. **L. hirtellus**, Greene, is also probably only a form of the same species. **L. formissimus**, Greene, is **H. gracilis**; **L. pinnatus** is **H. bicolor**. **L. eriophorus**, Greene, is **H. tomentosa**. **L. leucophæus**, Greene, is distinguished from Bentham's **H. grandiflora** by velvety instead of nearly glabrous leaves and smaller flowers; perhaps better considered as **Hosackia grandiflora**, Benth., var. **anthylloides**. Gray. **L. Benthami**, Greene, is **H. cytisoides**.

L. Bioletti, Greene. Slender, wiry branches, 2 feet long, prostrate, canescent, with short oppressed hairs: leaflets usually 4, cuneate-obovate: umbels pedunculate, bracteate, 6–10 flowered: calyx-teeth triangular, erect: corolla 2 lines long. We would label this **H. Bioletii** (Greene).

LATHYRUS.

L. Jepsonii, Greene. Nearly or quite glabrous: stem 5–8 feet long, strongly winged: leaflets 8–12, linear lanceolate, acute, 2–3 inches long: stipules small, setaceously acuminate: peduncles about equaling leaves: flowers rose-purple: calyx teeth ovate lanceolate, nearly equal in length: pod 12–16 seeded.

L. puberulus, White. Low, or sometimes 8–15 feet high and shrubby at the base, soft, hairy, or nearly glabrous, the stems angled; leaflets 5–7 pairs, ovate-oblong to linear, cuspidate: flowers purplish.

VICIA.

V. Hassei, Watson. Like **V. exigua**, but leaflets deeply notched: pods 5–8 seeded.

CERCIS and PICKERINGIA.

Siliquastrum, in *Bay-Reg. Bot.*, is **Cercis**, and **Xylothermia** is **Pickeringia**.

AMORPHA.

A. hispidula, Greene, differs from **A. Californica** in more numerous leaflets (17–25) which are retuse or emarginate. This, according to Greene, is the species from Monterey northward.

ROSACEÆ.

Herbs, shrubs or trees, with alternate leaves, usually evident stipules, mostly numerous stamens borne on the calyx; distinct free pistils from one to many, or in one suborder few and coherent with each other and adherent to the calyx forming a 2-several celled inferior ovary.

Nearly all the cultivated fruits of the temperate zones belong to this order. Key to genera and species, p. 113.

In *Bay-Reg. Bot.* all but two of the genus **Prunus** are put under **Cerasus**, with the same specific names. **Osmaronia**, Greene, is **Nuttallia**, Gray; **Opulaster capitatus**, Greene, is **Physocarpus opulifolia**, Max. **Malus** in Greene's book is **Pirus** in this. **Cercocarpus betulæfolius** is **C. parvifolius**. **Rubus parviflorus** is **R. Nutkanus; R. vitifolius** is **R. ursinus**.

HOLODISCUS.

There are probably two species, as given in the *Bay-Reg. Bot.*:

H. discolor, Max. Shrub, 2–6 feet high, branches ridged: leaves ovate, narrowed to a short petiole: panicles erect.

H. ariæfolius, Greene. Shrub, 8–18 feet high, with spreading or recurved branches bearing drooping panicles 6–10 inches long; leaves pinnately shallow-lobed from base to apex.

POTENTILLA.

In *Flora Franciscana* and *Bay-Reg. Bot.*, Prof. Greene has, with apparently good reason, united the genera **Horkelia, Ivesia, Sibaldia** and **Potentilla** under the latter name.

P. millegrana, Englm. Next to **P. glandulosa.** Tall, flaccid, soft-hairy: leaves 3-foliolate: minute flowers, numerous, yellow: stamens usually 10. Lower San Joaquin.

P. biennis, Greene. Biennial: stems erect, purple: leaflets 3, fan-shaped, irregularly incised: cymes of small yellow flowers dense: stamens about 10.

P. frondosa, Greene. 3 feet high, viscidly hairy and heavy-scented: leaflets 5–9, doubly incised, thin: bractlets long as calyx lobes or longer, trifid: flowers white. Martinez. *Frank Swett.*

P. Californica, Greene. Like the last, but leaves mostly radical: glandular, hairy, fragrant: leaflets 11-21, broadly wedge-form, and incised at apex: bractlets usually 3-toothed, exceeding the calyx-lobes: flowers white.

P. Parryi, Greene: slender stems, 6–10 inches high: leaflets many, cleft scarcely to the middle: calyx rotate, lobes longer than the narrow bracts: petals 3 lines long, white. Ione.

P. laciflora, Drew. Leaflets divided into 2 or 3 segments, bractlets narrow and much shorter than the calyx-lobes: flowers white: filaments petaloid-dilated: achenes only 2 or 3. Hy-am-pum, Trinity River.

P. Michneri, Greene. (*Horkelia.*) Leaflets about 15 pairs: all 10 stamens, with oblong-petaloid white filaments. Mt. Tamalpais.

P. Kelloggii, in *Bay-Reg. Bot.,* is **Horkelia Californica,** var. **sericia**; **P. tenuiloba** is **H. tenuiloba.**

Roses are sensitive to the influence of their homes, and prone to variation; hence botanists will never agree as to the number and limitation of species. The dwarf roses of Sonoma County and Mt. Tamalpais (**R. Sonomensis,** Greene) only a foot or less in height, with numerous small flowers, may be popularly known, at least, as the *Sonoma Rose;* and the dwarf rose (**Rosa spithamia,** Wat.), so abundant on the Trinity River and in the northern Sierra Nevada, may be called the *Span-high Rose.* **R. gratissima,** Greene, a form of **R. Californica,** distinguished by the fragrance of its leaves, may be called the *Kern River Sweetbriar Rose.*

CALYCANTHACEÆ.

Fragrant shrubs, with opposite, entire, extipulate leaves, and solitary, terminal, large dull-red or purplish flowers: numerous sepals and petals, all colored alike: stamens many: pistils many. **Butneria Occidentalis,** Greene, is **Calycanthus.** See p. 118.

SAXIFRAGACEÆ.

Herbs, shrubs, or small trees, distinguished from *Rosaceæ* by albuminous seeds; usually by definite stamens, not more than twice the number of the calyx-lobes; commonly by

SAXIFRAGACEÆ. **37**

the want of stipules; sometimes by the leaves being opposite; and in most by the partial or complete union of the 2 to 5 carpels into a compound ovary. Seeds usually indefinite or numerous. Petals and stamens on the calyx. Styles inclined to be distinct. Only the *Hydrangieæ* have many stamens. Key to genera and species, p. 119.

SAXIFRIGA.

S. Marshallii, Greene. Perennial. Leaves radical; oblong, dentate: scape about a foot high, loosely panniculate: white petals with a pair of green spots: filaments club-shaped. Hoopa Val., on Trinity Riv., C. C. Marshall. Rogue Riv., Howell.

S. Californica, Greene, is **S. Virginiensis**, Michx.

BOYKINIA.

Therofon elatum, Greene, is **Boykinia occidentalis**, T. & G.

TELLIMA.

T. scabrella, Greene. A foot high, bearing bulblets: calyx with rounded base: petals entire, the two upper shorter and broader than the others: styles glabrous. Sierra Nevada.

MITELLA.

M. diversifolia, Greene (Marshall's Mitella). Leaf margins ciliolate: scape a foot high: shallow calyx-lobes, whitish petals palmately trifid at the abruptly widened apex: stamens 5. Trinity Mts.

RIBES.

Ribes tenuiflorum, Lindley, is, according to Prof. Greene, our species of *Golden Currant*, not **R. aureum**.

R. ambiguum, Wats. Glandular and hairy: spines short: flowers mostly solitary, 6 lines long or less, greenish: stamens about equaling the white petals: small anthers, obtuse: fruit densely spinose. Trinity River, northward.

R. Marshallii, Greene. Glabrous: flowers an inch long: calyx dark purple: petals 2-3 lines long, salmon color. Trinity Mts. May be a variety of the last.

R. Victoris, Greene. (Chestnut's Gooseberry). 5 feet high: spines triple: leaves viscid: greenish flowers, 6 lines long: calyx-tube much exceeding the lobes, petals white, acute, and toothed at the apex: anthers sub-sagittate, mucronate: ovary glandular, rough-hairy. Marin and Napa Cos.

R. Californicum, H. & A. Rigid, flexuose, glabrous branches: leaves small: petals thick, truncate, erose-toothed at the end: stout filaments, three times the length of the petals: berry prickly.

R. subvestitum, H. & A. Taller, 5-10 feet high: leaves very viscid and heavy-scented: petals truncate, entire.

R. amictum, Greene, is the Sierra Nevada variety of **R. Menziesii,** which Prof. Greene thinks is confined to the Coast Region. All these prickly-fruited Ribes may be considered varieties of *Menzies' Gooseberry.*

CRASSULACEÆ.

Fleshy plants, with sepals, petals, stamens, and distinct carpels of the same number (3–12), or the stamens twice as many: polypetalous or gamopetalous. Key to genera and species, p. 122.

TILLÆA.

T. Drummondii, T. & G. and *T. Bolanderi,* of *Bay-Reg. Bot.,* may be considered varieties of **T. angustifolia.**

DROSERACEÆ.

Low bog herbs, purplish or brownish, with radical leaves, bristly with gland-tipped hairs which secrete a viscous fluid. Flowers in, usually, scorpioid racemes or spikes: calyx 5-parted: petals and stamens 5: styles mostly 3, each 2-parted. The most remarkable insectivorous plants belong to this order. Key to genera and species, p. 123.

LYTHRACEÆ.

Our species, herbs with entire leaves. Flowers with tubular calyx, bearing the petals and stamens on its throat, and rather closely inclosing the superior ovary: style one. Key to genera and species, p. 124.

AMMANIA.

A. coccinea, in *Bay-Reg. Bot.,* is **A. latifolia,** and **Rotala ramosior** is **A. humilis.**

LYTHRUM.

L. Sanfordi, Greene. Erect stem, acutely 5–6 angled: petals 6, bright purple. Much like **L. Californicum,** of which it may be a variety.

L. adsurgens, Greene. Branches 5-angled, 1–3 feet long, decumbent or assurgent, slightly succulent: calyx 2½ lines long, 12-striate: petals pale purple. Similar to **L. hyssopifolia,** but perennial instead of annual, and much larger.

HALORAGEÆ.

Aquatic herbs, with inconspicuous, often apetalous flowers, sessile in the axils of leaves or bracts: calyx adherent to the ovary in the fertile ones, and its lobes then short or obsolete. Flowers perfect, but apetalous, in *Hippuris,* and monœcious or perfect in *Myriophyllum.* Key to genera and species, p. 124.

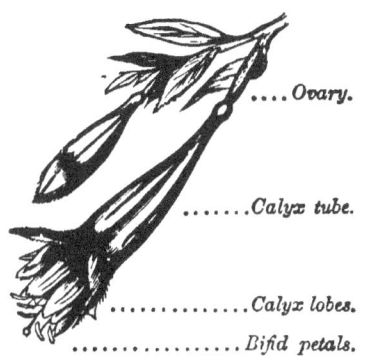

.... *Ovary.*

...... *Calyx tube.*

.... *Calyx lobes.*

.... *Bifid petals.*

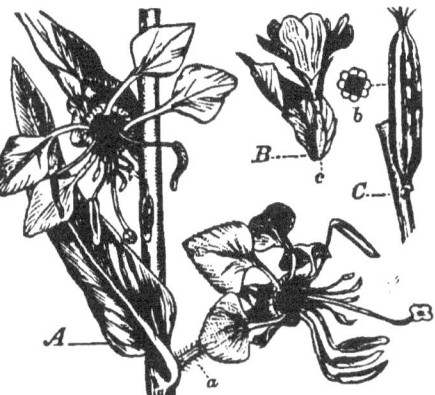

ONAGRACEÆ.

Herbs (shrubby exotics), with the parts of the flowers usually in fours, the calyx tube adnate to the ovary, the petals borne on its throat, and the stamens as many or twice as many. Style always single. In **Jussiæa** the flower parts vary in number from 4 to 6; in **Circæa** the parts are in twos.

Fig. A. Clarkia elegans; *a*, inferior, sessile ovary of the axillary flower. Fig. B. Boisduvallia densiflora; *c*, inferior ovary, sessile in the axil of a bract. Fig. C Capsule of Godetia; *b*, cross section of the same. Fig. D. Epilobium paniculatum; *h*, inferior ovary; *f*, a grown capsule; *g*, tube of calyx above the ovary; *e*, one of the bifid petals; *i*, one of the seeds bearing a tuft of silken hairs. The figure on the left is a common form of Zauschneria.

Many of our plants blossom late in the dry season. These usually have hard stems from which a shedding, thin, outer bark hangs in shreds. Key to genera and species, p. 124.

JUSSIÆA.

J. repens, L., var. **Californica,** is **J. diffusa,** Forsk., in *Bay-Reg. Bot.* It grows in floating masses on stagnant water or along the edges of ponds.

LUDWIGIA.

L. palustris, Linn., is **Isnardia palustris** in *Bay-Reg. Bot.* Leaves oval or ovate, acute, 6–12 lines long: sessile flowers, solitary in the axils: petals when present, minute. Muddy borders of ponds or watercourses.

ZAUSCHNERIA.

Z. Californica, Presl. Decumbent stems, often many together, from a woody base, branching, more or less hairy: leaves ovate to lanceolate. Blossoming from June to December. (See the figure on the left above.)

Prof. Greene makes several species of this variable plant.

EPILOBIUM.

E. rigidum, Haussknecht. Cespitose perennial, a span or two high: leaves lanceolate to obovate, acute, entire, often oblique, narrowed into short-winged petioles, very glaucous; flowers in the axils of the small upper leaves which are often adnate to the bases of the peduncles: ovary densely white-glandular: petals 7–10 lines long: stigma very large. S. W. Oregon, as is also the var. **canescens**, which is densely velvety-canescent.

Epilobium paniculatum.

E. palustre, Linn. Perennial, a foot or less high, canescent above with incurved hairs: leaves mostly opposite, evidently veined, narrowly oblong, obtuse, or almost truncate: flowers small, the calyx-tube funnelform: stigma club-shaped. State of Washington.

E. adenocaulon, Hausskn. Perennial. Branches upcurving, glandular-hairy above; leaves elliptical to ovate-lanceolate, slightly serrulate, rather pale green and glossy: flowers 2 or 3 lines broad: stigma club-shaped.

E. Californicum, Hausskn. Taller than the preceding, inflorescence and buds white, with long ascending hairs: leaves often 3 or 4 inches long, lanceolate, remotely serrulate: flowers few, 2 or 3 lines broad: slender fruiting peduncles sometimes equaling the leaves.

E. brevistylum, Barbay. Slender, a foot high: leaves opposite, ovate or elliptical, an inch long or more: flowers 2 lines broad: pods 2 in. long.

E. ursinum, Parish. A foot high or less, slender, hairy below, minutely glandular above: leaves mostly opposite, about an inch long, ovate to broadly lanceolate, serrate: flowers white or lavender, 2 lines broad: pods 15 lines long on peduncles half as long.

E. Hornemanni, Reich. A span or two high, slightly hairy above; leaves an inch long, elliptical, ovate, obtuse: flower 3 or 4 lines broad, lilac to violet: pods about 2 inches long on peduncles equaling the subtending leaves.

E. Oregonense, Hausskn. A span high, with flowerless shoots at the base: leaves 8–12 lines long, crowded below, remote, and very small above, oblong to linear obtuse: flowers 3 lines broad, violet: pods 2 inches long on peduncles nearly as long.

E. clavatum, Trelease. A span high, densely cespitose, glabrous: leaves divergent: 5–10 lines long, broadly ovate, very obtuse: flowers 2 lines broad, rose color: pods an in. long.

E. holosericeum, Trel. Rather woody, loosely branched, upper leaves and branches canescent, with sub-appressed hairs: leaves oblong lanceolate, remotely serrulate; flowers pale, 2 lines broad: pods 2½ inches long on peduncles 6 lines long.

GAYOPTYUM.

G. lasiospermum, Greene. Erect, very slender, 1 or 2 feet high, with numerous dichotomous branches: corolla 1½ lines long, rose color: seeds canescent, with appressed silky hairs (hence the name). San Diego Co., northward.

EULOBUS.

Œnothera leptocarpa, in *Bay-Reg. Bot.*, is Eulobus Californica. Southern Coast.

ŒNOTHERA.

Œ. Hookeri, T. & G. Biennial: stem red, stout, angular, 3-6 feet high: petals nearly 1½ inches long, obcordate, pale yellow turning to rose color. Common in the southern counties. Probably a variety of *Œ. biennis.*

Œ. grandiflora, Ait. Differing from *Œ. biennis* by its larger, almost scentless flowers, declined stamens, stigma lobes yellow, not green. This may be considered the cultivated form of *Œ. biennis.*

Œ. arguta, Greene. Perennial, stems slender, about a foot high: leaves linear-lanceolate, dentate, 2-4 inches long, sessile: calyx tube 1½ inches long: petals as long, deeply obcordate, bright yellow turning to orange. Moist ground near Monterey and southward.

Œ. nitida, Greene (§ 3). Decumbent branches, a foot long or less, very rigid: leaves spatulate or oblanceolate, rather fleshy, dark, lustrous green: petals 6 lines long: anthers linear-oblong, fixed, almost in the middle: capsule 10 lines long, acutely angled. Monterey Bay, Abbott.

Œ. hirtella, Greene (§ 3). Erect, simple, or branches from the base, 6-10 inches high, purplish, short hairy: radical leaves, oblanceolate, denticulate, 1½ inches long: stem leaves ovate, sessile, ½ inch long, coarsely toothed and more or less crisped: petals a line long or more: capsule hairy, attenuate upward, once or twice coiled. Interior hills.

Œ. spiralis, in *Bay-Reg. Bot.*, is **Œ. cheiranthifolia** and **Œ. campestris** is **Œ. dentata. Œ. gauræfolia** has white or pink flowers half au inch broad.

Œnothera ovata.
S—Surface of ground.
r—Rootstock.

GODETIA.

G. micropetala, Greene. Erect, 1–3 feet high: spike rather short: calyx tube 2 lines long, segments 4 lines, the slender tips twisted in the bud: petals linear-lanceolate, 3 lines long: stigma purple: capsule sessile. Contra Costa Co.

G. rubicunda, of the *Bay-Reg. Bot.*, may be a variety of *G. Amœna.*

CLARKIA and EUCHARIDIUM.

In *Bay-Reg. Bot.*, **Eucharidium** is put under **Clarkia.** **C. grandiflora** is probably a more branching and larger flowered form of **Eucharidium concinnum.** **E. Breweri** is the rarest and most beautiful of the genus. It was discovered on Mt. Oso in Stanislaus Co., and has been collected at the Geysers and on Mt. Hamilton. The flowers have air almost perfectly square outline, the long middle lobes of the petals forming the angles.

BOISDUVALIA.

B. glabella, Walp. A span or two high: leaves ovate-lanceolate, serrulate, bluish, densely soft-hairy to glabrous, 5 lines long: flowers in a terminal, leafy spike, or some in the lower axils, about a line long.

LOASACEÆ.

Herbaceous plants with either stinging or jointed and rough-barbed hairs; no stipules, calyx tube adnate to the 1-celled ovary. Stamens usually very numerous. Key to genera and species, p. 128.

MENTZELIA.

Calyx cylindrical to ovoid; the persistent limb 5-toothed. Petals 5 or 10: stamens numerous, inserted below the petals on the throat of the calyx: filaments free or in clusters opposite the petals, filiform or the outer petaloid. Style 3-cleft, the lobes often twisted. The leaves are alternate, mostly coarsely toothed or pinnatifid; flowers white to yellow or orange.

M. affinis, Greene. Similar to *M. dispersa,* but stouter, often 2 feet high, simple and leafy below, widely branching above: leaves lanceolate, deeply sinuate-pinnatifid: flowers 6 lines broad: capsule an inch long, slender.

M. pectinata, Kell. Stem usually simple, 4–8 inches high, clothed like the leaves with barbed hairs: flowers deep yellow, an inch broad: petals mostly obcordate, with a minute cusp in the sinus: stamens numerous, half as long as the petals. Marysville Buttes southward.

L. lævicaulis is perhaps always found on the flood beds of streams. Its flowers have narrow petals, lacking the satiny luster which marks the other large flowered species.

CUCURBITACEÆ.

Tendril-bearing, trailing, or climbing herbs, usually rough and rather succulent. Flowers axillary to alternate leaves, solitary or clustered, monœcious. Calyx adherent to the ovary, the limb 5-lobed. Corolla with united petals. Stamens usually 3, united. Pistil 2–3 carpeled. Squashes and pumpkins, natives of America, with melons, cucumbers and gourds, natives of the Eastern Continent, are the common cultivated plants of this order. Key to genera and species, p. 129.

DATISCACEÆ.

In our genus, stout, glabrous, diœcious, perennial herbs, with laciniate-pinnatifid leaves, the greenish flowers clustered in the axils. Key to genera and species, p. 129.

CACTACEÆ.

Green, fleshy, and thickened, persistent (though mostly herb-like), usually leafless plants; globular or columnar, or jointed and often flattened, usually armed with bundles of spines from the axils of absent leaves. Flowers with numerous sepals, petals, and stamens, the cohering bases of all coating the inferior 1-celled ovary, and forming above it a tube or cup: style 1 with several stigmas. Key to genera and species, p. 129.

FICOIDEÆ.

Usually very succulent plants with opposite leaves. (In our plants *Mollugo* is not succulent, and *Tetragonia* has alternate leaves.) Petals and stamens various, the former often wanting: carpels 2 to many. Key to genera and species, p. 130

TETRAGONIA.

T. expansa, Murr., a native of New Zealand, and cultivated under the name of *New Zealand Spinach*, is apparently a native of our coast, growing on the shore of San Francisco Bay. It is described as follows in *Bay-Reg. Bot.* Perennial, with alternate, plane, fleshy leaves, and axillary, greenish, apetalous flowers: fruit 4-horned, about ½ inch long, scarcely as broad.

ARALIACEÆ.

Herbs, shrubs, or trees, with mostly stout, hollow stems, and alternate lobed or compound leaves. Flowers in simple umbels, which are often arranged paniculately or racemosely· calyx adhering to the ovary: petals 5-10: stamens as many or twice as many: ovary more than 2-celled.

A tall herb: leaves bipinnate or pinnate, very large: pedicels jointed............... 1

Stem woody, prickly, 6-12 feet high: leaves palmately lobed: pedicels not jointed.... 2

1. ARALIA, Linnæus.

1. **A. Californica**, Watson. Root large, aromatic, used medicinally. (*Spikenard.*)

2. FATSIA, Bentham & Hooker.

1. **F. horrida**, B. & H. Common in the forests of Oregon and northward.

UMBELLIFERÆ.

Herbs with small flowers in umbels, stamens and petals 5, borne on a 2-celled ovary which in fruit splits into a pair of dry, usually flat, indehiscent carpels. Since the generic distinctions depend upon characters of fruit and seed difficult of determination, the plants of this order are not here described.

CORNACEÆ.

Trees or shrubs, rarely herbs, with simple entire mainly opposite leaves, no stipules, and flowers in cymes, capitate clusters or spikes; the petals and stamens 4, epigynous; calyx adnate to the 1-2 celled ovary, which becomes a drupe or berry. Key to genera and species, p. 131.

CORNUS.

C. stolonifera, Michx. Stems numerous, clustered, decumbent, forming a low thicket: twigs nearly glabrous, red-purple: leaves mostly oval or oblong, acute: cymes small, flat-topped: fruit white, globose, stone furrowed on the edges. Trinity Mts., C. C. Marshall.

C. Baileyi, C. & E. Stone twice as long as high, flattened edge, furrowed. Castle Rock, Columbia River.

DIVISION 2. GAMOPETALÆ.

CAPRIFOLIACEÆ.

Shrubs, trees, woody climber or trailing plants. Leaves opposite: calyx adherent to the ovary, the limb 5-toothed or obsolete: corolla 4-5 lobed or cleft: ovary 2-5 celled: fruit a berry. Key to genera and species, p. 131.

SAMBUCUS.

S. maritima, of the *Bay-Reg. Bot.*, seems to be only a form of S. glauca; and S. callicarpa, Greene, is S. racemosa of this book, an Old World species, probably distinct from ours, making Greene's the better name for the Red-fruited Elder.

SYMPHORICARPOS.

S. ciliatus, Nutt., is the name given, in *Bay-Reg. Bot.*, to the small *Snowberry* of the Oakland Hills, which is perhaps only a variety of S. racemosus, the most common species, or, perhaps, the only one of the Coast Ranges.

LONICERA. (CAPRIFOLIUM, in Bay-Reg. Bot.)

L. interrupta, Benth., is distinguished from L. hispidula (in *Bay-Reg. Bot.*) by erect bushy habit, white bark of branches, and glabrous yellow flowers, smaller.

L. Ledebourii, Esch., is distinguished in *Flora Franciscana* from L. involucrata by larger size, often 10-15 feet high, while the latter is only 2 or 3 feet high: more gibbous corolla, salverform rather than funnelform, and orange to scarlet, instead of yellowish. Named Distegia Ledebourii in *Bay-Reg. Bot.*

L. subspicata, H. & A. Bushy, many branched, densely glandular-hairy, except on the upper side of the leaves, which are small, narrowly oblong, tapering to a petiole, none of them stipulate or connate: corolla 6 lines long, yellowish. Usually considered a variety of L. *hispidula*.

RUBIACEÆ.

Known by having opposite entire leaves with intervening stipules, or whorled leaves without stipules, along with an inferior ovary and regular 4-5 merous flowers; the teeth of the calyx sometimes wanting. Stamens alternate with the lobes of the corolla and borne on its tube, distinct. Key to genera and species, p. 133.

Sherardia arvensis, Linn., an introduced weed, called in England Field Madder, has been found at Berkeley by Professor Greene. A small annual, about 6 inches high, bearing leaves in whorls of 6: flowers small, blue or pink, in little terminal heads surrounded by a broad, deeply lobed involucre.

GALIUM.

G. spurium, Linn. Distinguished from **G. aparine** by pale green flowers and pedicels recurved in fruit.

G. Anglicum, Huds. Distinguished from these by greenish flowers followed by fruit, not rough, with hooked hairs. All three introduced from the Old World. Perhaps better considered varieties of **G. Aparine**.

VALERIANACEÆ.

Herbs with opposite leaves, no stipules: the distinct stamens fewer than the lobes of the corolla, and borne on its tube: the inferior ovary with two empty cells, and one containing a solitary ovule, ripening into a kind of akene. Key to genera and species, p. 133.

DIPSACACEÆ.

Herbs with opposite leaves and flowers in dense involucrate heads, each flower inclosed in a tubular involucel, and subtended by a bract. Calyx adherent to the ovary: corolla 4–5 lobed, bearing stamens alternate with the lobes: fruit crowned with the calyx limb, 1 seeded. Key to genera and species, p. 134.

Scabiosa atropurpurea, Linn., has run wild in some localities. It may be known by its pinnate leaves and heads of black-purple or lighter colored flowers, even white, the outer circles larger, and the calyx in fruit stem-like above the akenes, with 5 spreading bristle-like lobes. Commonly called *Mourning Bride*.

COMPOSITÆ.

Flowers, usually many in a dense head, sessile, on a common receptacle, surrounded by a calyx-like involucre: the calyx reduced to hairs or scales, or obsolete: the corolla tubular, equally lobed, ligulate or bilabiate, the 5 stamens united by their anthers into a tube inclosing the 2-parted style: the ovary inferior, forming in fruit an akene which is usually crowned with the persistent calyx (pappus).

Sunflowers, marigolds, thistles, and dandelions are types of the conspicuous plants in this order. This, the largest of all the orders, is represented in California by over 500 species. Although the flower heads are frequently large, the separate flowers, with but few exceptions, are too small to be examined without the aid of a microscope skill-fully used. The order is, therefore, far too difficult for the beginner.

LOBELIACEÆ.

Herbs, mostly with milky juice, alternate simple leaves, scattered or racemose flowers: the calyx adherent to all or half of the ovary: the corolla more or less irregular: the stamens united into a tube closely inclosing the style: stigma commonly 2-lobed. Key to genera and species, p. 134.

HOWELLIA, Gray.

H. aquatilis, Gr. Aquatic: submersed leaves, slender, mostly alternate, entire; those above water broader and shorter, sometimes 1-2 toothed: flowers short-peduncled, axillary: corolla lobes nearly equal, not surpassing the calyx. Ponds on Sauvies Island, Columbia River.

H. limosa, Greene. Weak procumbent branches, 6-12 inches long, leafy and florif-erous throughout: leaves lanceolate, entire, sessile, 6 lines long: corolla white, the wedge-shaped divisions a line long, the two upper narrower. On the margins of pools near Suisun.

DOWNINGIA.

In *Bay-Reg. Bot.*, the following species are described under the generic name **Bolelia**:

D. insignis, Greene. Erect, mostly simple, stems, few flowered: lower lip of the corolla 6 lines broad, obovate, 3-lobed, the lobes and sides sky blue, darker veined, the main portion white, bearing in the middle two oblong green spots: upper lip merely bifid, the lobes ascending and parallel: throat of corolla with a pair of yellow folds in a field of dark violet.

D. tricolor, Greene. Branches weak and reclining: lower lip of corolla parted into 3 equal, broadly obovate, truncate, and slightly cuspidate lobes, these deep blue at tip, white below, the undivided part with a transverse, somewhat quadrate spot of dark maroon: upper lip of two small segments slightly recurved, parallel.

D. concolor, Greene. Numerous branches, forming a dense tuft: corolla all violet, but base of lower lip very dark, bordered by lighter blue, the lobes slightly unequal, very obtuse and somewhat cuspidate, upper lip cleft to the middle only.

D. ornatissima, Greene. Taller than the preceding, 6-10 inches high: corolla pale, 4 fold-like protuberances partly filling the throat: segments of the upper lip coiled backward into a ring, the corolla tube at base of these segments abruptly raised into a sharp protuberance.

D. humilis, Greene. Only an inch high: segments of the calyx unequal, exceeding the minute white corolla.

D. cuspidata (Greene). [Determined since the publication of *Bay-Reg. Bot.*] Erect, slender, 6 inches high or more, with few and small leaves and few remote flowers: lower lip of corolla nearly 6 lines broad, only 4 lines deep, the lobes broadly ovate, retuse, or obcordate with a cusp, the terminal half violet, the base white; the undivided part of the lip yellow, without folds, protuberances or depressions: lobes of the upper lip 1½ lines long, cuspidate, straight, violet. Sonoma and Napa Co's.

CAMPANULACEÆ.

Herbs with alternate leaves without stipules and regular flowers, having the calyx adnate to the ovary, distinct stamens (5, rarely 4) inserted with the corolla, alternate with its lobes: calyx persistent. Stamens with introrse anthers, opening in the bud. Style single, its upper portion beset with hairs which collect the pollen, its summit 2–5-lobed or cleft. Key to genera and species, p. 135.

Specularia is **Triodanis**, in *Bay-Reg. Bot.*

GITHOPSIS.

G. diffusa, Gray. Slender, diffusely branching, glabrous: calyx-lobes subulate-lanceolate, half the length of the slender sessile pod. Vaca Mts. and southward.

ERICACEÆ.

Woody plants or perennial herbs, with symmetrical and mostly regular flowers: the stamens as many or twice as many as the petals or lobes of the corolla, and inserted with but rarely upon it: the anthers 2-celled, and the cells opening by a terminal pore; the ovary with as many cells as the divisions of the corolla or calyx: the seeds small. Corolla generally gamopetalous, sometimes of distinct petals, the insertion and that of the stamens hypogynous, or when the calyx is adnate epigynous around an annular disk. Style single. Leaves simple. Key to genera and species, p. 136.

ARCTOSTAPHYLOS.

A. myrtifolia, Parry. Widely branching from the base, 1–3 feet high: leaves entire, ovate, 4–10 lines long, 3–4 lines wide, acute at both ends with a thickened margin. Growing in patches on gravelly ridges east of Ione, Amador Co.

A. manzanita, Parry. Varying in size from a small shrub to a tree, 25 feet high, with a trunk a foot in diameter: leaves petiolate from narrowly to broadly ovate, usually obtuse: pedicels smooth: calyx lobes orbicular: corolla broad: filaments slightly hairy: fruit 4–6 lines broad, 3 lines high, changing from dull white to red-brown: early flowering, often in bloom in the Bay counties on Christmas. According to Parry, this is the species which has heretofore been called **A. pungens** which is a Mexican species.

A. viscida, Parry. Distinguished from **A. glauca** by very viscid pedicels which are four or five times as long as the bracts which are also viscid: flowers light pink: fruit flattened, 3 lines broad, 2 lines high. Foothills of the Sierra Nevada from Central California northward.

A. Stanfordiana, Parry. Low, branching, 3–5 feet high: leaves bright green on both sides, narrowly ovate to oblanceolate, tapering into a narrowly winged petiole: inflorescence smooth: calyx deep red: corolla pink: style becoming exserted. Near Calistoga, Napa Co. Named for *Leland Stanford, Jr.*

RHODODENDRON.

R. Somonense, Greene. Shrub 2–5 feet high; leaves nearly elliptical, 1 in. long or less, the margin serrulate and ciliolate: flowers rose-color, an inch long or more. (Perhaps a var. of *R. occidentale.*) Sonoma Co. (Hence the name.)

Gaultheria is **Brossæa,** in *Bay-Reg. Bot.*

NEWBERRYA.

According to rules of nomenclature likely to prevail this generic name must give place to **Hemitotes,** which Gray first gave to the plant discovered by Newberry. **H. pumilum,** described by Greene, is exactly **N. congesta** of this Key. (Collected in Mendocino Co. by W. G. Wright.)

LENNOACEÆ.

Fleshy root-parasites. Parts of the flower almost always more than 5: stamens adherent up to the throat of the tubular corolla. Key to genera and species, p. 140.

PLUMBAGINACEÆ.

Seashore herbs, with radical leaves. Flowers regular, all the parts in fives: calyx 5-plaited, 5-toothed, persistent: petals with long claws united into a ring at the base: Stamens opposite the petals. Key to genera and species, p. 140.

Statice Limonium, var. **Californica,** is **Limonium commune,** var. **Californicum,** in *Bay-Reg. Bot.,* and **Armeria vulgaris** is **Statice Armeria.**

PRIMULACEÆ.

Herbs, with perfect, regular flowers, well marked, by having the stamens as long as the lobes of the corolla, and opposite to them, inserted on its tube: a single entire style

Botany — 4

and stigma, a one-celled ovary, and capsular fruit. Calyx 4–8 cleft, commonly 5-cleft, hypogynous: leaves simple: stipules none. In *Glaux* the corolla is wanting: stamens on the calyx, alternate with its lobes. Key to genera and species, p. 141.

DODECATHEON.

Species of this genus are described under the generic name **Meadia**, in *Bay-Reg. Bot.* and the variety with cream-colored flowers is there named **Meadia patula**. The many, forms of this genus may well bear one specific name.

ANDROSACE, Matthiolus.

A. acuta, Greene. Very slender, 1–4 inches high, rough-hairy: leaves radical, linear-lanceolate, entire, 6–9 lines long: flowers in an involucrate umbel on a scape: corolla salverform, white.
Trientalis is **Alsinanthemum** in *Bay-Reg. Bot*

STYRACACEÆ.

Shrubs or trees with alternate, simple leaves, a calyx adherent at least to the base of the ovary: stamens usually at least twice the number of corolla lobes and more or less united to each other and to the corolla. In our species the white downy flowers are about an inch broad, and the filaments are united nearly half their length. Key to genera and species, p. 142.

OLEACEÆ.

This, the *Olive Family*, is represented on the Pacific Coast by two ash trees, which may be known by opposite pinnate leaves, leaflets 5–9, oval to oblong, the fruit a winged akene. Key to genera and species, p. 142.

APOCYNACEÆ.

Herbs with milky juice, opposite entire leaves, and regular 5-merous flowers. Ovaries 2, but stigmas united and the anthers adherent. Seeds in our species bearing a tuft of silky down at the end. Key to genera and species, p. 142.

APOCYNUM.

A. pumilum, Greene. Commonly hairy: lowest leaves subreniform to round-ovate, ovate, the others cordate-ovate and oval, rarely more than an inch long: corolla subcylindrical, 3–4 lines long, the segments scarcely spreading.   Monte Diablo Range and northward.

ASCLEPIADACEÆ.

Herbs with milky juice, no stipules, and regular flowers, with the parts in fives, except that there are two carpels with distinct ovaries and a common stigma to which the stamens are attached; the latter (in our genera) with hood-like appendages: leaves entire, generally opposite, sometimes whorled: flowers usually in simple umbels: fruit a pair of follicles.   Seeds almost always with a coma of silky down.   Key to genera and species, p. 142.

ASCLEPIAS.

In *Bay-Reg. Bot.*, **A. Californica**, Greene, is **Gomphocarpus tomentosus**, of this book, **A. ecornuta** is **Gomphocarpus cordifolius**, and instead of **Schiznotus**, Greene writes **Solanoa**.   The latter plant grows near the Geysers of Sonoma Co.— where it was first collected by *C. B. Towle*—and on the mountains north of Clear Lake. Its red flowers and often prostrate habit make it a noticeable plant in the order.   It may well be called *Towle's Milkweed.*

GENTIANACEÆ.

Glabrous herbs, with colorless, bitter juice, entire, opposite and sessile leaves: no stipules, perfect and regular flowers: stamens as many as the lobes of the corolla and alternate with them, inserted on the tube, the anthers free from the stigma: ovary 1-celled: style one or none: the stigmas commonly two.   Calyx persistent.   Key to genera and species, p. 143.

POLEMONIACEÆ.

Chiefly herbs with simple or divided leaves, and no stipules: all the parts of the regular flower five, except the pistil, which has a 3-celled ovary and a 3-lobed style.   Calyx imbricated in the bud, persistent: corolla convolute in the bud: stamens on the corolla alternate with its lobes, distinct: anthers introrse.   In *Gilia* the cells of the ovary and the stigmas are occasionally reduced to two.   Key to genera and species, p. 145.

It is very difficult to define the genera of this order.   If we arrange all our species in groups according to their affinities these groups interlace more or less deeply.   In other words, one or more species are common to two or more groups.   Therefore, when we separate these groups under generic names there are species that might as well be put in one as the other of adjacent genera.   Before many of these troublesome intermediate species (connecting links) were known, botanists easily made out a dozen or more genera. When the first edition of this Key was published in 1886, Dr. Asa Gray reduced the genera to three. Since then Prof. Greene has grouped our species under six genera in his *Manual of Bay Region Botany*.

GILIA.

In *Bay-Reg. Bot.*, sections *Dactylophyllum*, *Linanthus* and *Leptosiphon* of this book form the genus **Linanthus**, in which **L. filipes** is **G. pusilla**, var. **Californica**; **L. grandiflorus** is **G. densiflora**; **L. bicolor** is **G. tenella**, and three, perhaps good species, forms that have usually been put with **G. androsacea**, viz: **L. parviflorus, L. acicularis**, and **L. rosaceus**; the first a common slender form with pale yellow or white corolla, tinged outside with red or brown; the second with golden yellow flowers, and the third, the rather stout short form with rose-red flowers, so common on the San Francisco peninsula.

The following new species, not of the Bay Region, should appear under **Gilia**, as follows:

G. serrulata, Greene. Freely and almost diffusely branched, 6 or 8 inches high; leaf-segments and floral bracts all linear-acerose, the margins spinulose-serrate: calyx segments more than twice the length of the tube: corolla with slender, dark-purple, not far exserted tube, and narrowly funnelform throat, the limb of oblong-spatulate white segments, 9 lines broad.

G. montana, Greene. Habit of *G. ciliata*, but larger, less hispid: corolla much larger, nearly 2 inches long, the tube widening to a broadly funnelform purple throat: segments of the limb cuneate-obovate, truncate whitish.

G. nudata, Greene. Slender, 3–10 inches high, branches few: leaves 3 lines long, distant: flowers in dense clusters: corolla tube long, exserted, short yellow throat, and white or purplish limb, 6 lines broad. Lake Co.

In *Bay-Reg. Bot.*, the species of § **Navarretia** are put in the genus **Navarritia**, to which are added two new species—N. **parvula**, Greene, and N. **mellita**, Greene These would here be described under **Gilia**, viz:

G. parvula, Greene. Branching, 2–4 inches high, very viscid and aromatic: lowest leaves linear, entire: the upper broader, with teeth or segments: corolla about 4 lines long, broadly tubular-funnelform, light blue: stamens very unequal, the 2 posterior included, the 3 anterior long exserted and declined.

G. mellita, Greene. Slender, 2–5 inches high, very viscid and honey-scented (hence the name): lowest leaves pinnately divided into spine-like segments, those of the upper broader but spine-tipped: corolla minute, not exceeding the calyx, bluish: stamens included. Belmont, San Mateo Co. Summer blooming.

To the preceding may be added the following, published under the name **Navarretia**, not in *Bay-Reg. Bot.*:

G. **nigellæformis**, Greene. Habit of **G. cotulæfolia**: flower clusters conspicuously involucrate, the bracts broad and divided into bristle-like segments: 2 larger calyx-lobes bristly pinnatifid, the others 3, with pungent teeth: corolla deep yellow, with 5 purple or crimson spots in the funnelform throat.   Near Visalia, Dr. T. J. Patterson.

G. **mitracarpa**, Greene.   Depressed and diffuse, hairy,. the inflorescence glandular: leaves rigid and pungent, with about 2 pairs of 2-parted basal segments, and a lanceolate toothed terminal one: 2 calyx-lobes with a spinulose tooth on each side, 3 entire and shorter: pod 1-seeded.   Lake Co.

G. **prolifera**, Greene.   Erect, spreading, a span to a foot high, with rather large capitate flower clusters, the slender, naked, wiry branches radiating from beneath the earlier clusters and ending in similar heads: leaves an inch long, glabrous, linear-filiform, entire, or with one or two pairs of segments at the base: calyx and pungent bracts whitish with a viscid wool: calyx-tubes thin-membranous, longer than the teeth: corolla almost salverform, purplish or blue with a yellow throat.

G. **tagetina**, Greene.   Stems mostly strict and simple, a foot or more high, sparingly leafy, glabrous, glandless: leaves pinnately parted into 7–9 linear segments which are spinulose-toothed or pinnatifid: bracts divided into rigid pungent lobes, whitish hairy below: calyx segments very unequal, the 2 larger pinnately, the 3 smaller nearly palmately parted into ridged filiform divisions: corolla very slender, 9 lines long: ovules many.

G. **foliacea**, Greene.   Similar to *G. atractyloides*, but odorless, more diffuse and leafy, less spinose, the calyx-lobes very unequal: corolla white, little surpassing the calyx.

Under **Navarretia** Greene describes **G.** cotulæfolia of this book, as "rigid, 4–8 inches high, somewhat glandular: leaves twice pinnatifid into slender, herbaceous, soft and innocuous segments, the uppermost ones and the bracts decidedly spinescent: flowers white, 4-merous. A peculiar soft-leaved and scentless species." Greene also decides that **G. pubescens**, Benth., which has been confounded with **G. cotulæfolia**, Benth., is a good species, which may be distinguished by leaf segments, all with sharp and stiff teeth or lobes: calyx teeth all pungent, 3 small and entire, 2 twice as large and toothed: corolla deep blue or purple, 9 lines long: stamens exserted: odor strong, goatlike. **G. viscidula**, Gr., var. **heterodoxa**, Gr., is considered a good species by Greene, and is thus distinguished. Extremely viscid, the odor like that of a skunk: stamens exserted and declined.

Gilia in *Bay-Reg. Bot.* contains only species found in this book under § 7, **Hugelia**, and § 9, **Eugelia**. **G. gilioides** is **G. glutinosa** of this book. **Collomia** includes one each of § 11, **Courtoisia**, and § 12, **Collomia**. **G. graciles** of the last section is **Phlox gracilis** in Greene's book.

G. **leptalea**, Greene.   Distinguished from **G. capillaris** by being less glandular, less leafy and the leaves narrower, more slender and divergent branches and a much larger corolla, fully 6 lines long of a rich red-purple, while that of the latter species with which it has been confused is barely 2 lines long and white or pale purple. — These species were first clearly distinguished by Prof. Greene in Erythea, Mar. 1896.

G. millefoliata, Fisch & Mey., is according to Greene distinguished from G. multi-caulis most obviously by the corolla, that of the latter dark violet throughout and much larger than the 2-colored corolla of the former, which has white or bluish lobes, the throat with 5 dark spots.

G. abrotanifolia, Nutt. Nearly glabrous, 1–2 ft. high, simple or a few branches, these and the upper main stem naked and pedunculiform bearing a terminal dense cymose cluster of large blue flowers: calyx membranous except the midribs: stamens scarcely exserted. Santa Inez Mts.

G. Chamissonis, Greene, should, if the author of the name is correct, displace G. achillæfolia of this book and Bay-Reg. Bot., because the latter name was given by Bentham to a large form of G. multicaulis, while the plant heretofore known under that name was named Polemonium capitatum by Eschscholtz.

G. staminea, Greene must according to the above take the place of G. capitata.

G. Rawsoniana, (Greene) stems clustered from a perennial root, a foot high or more, sparingly branched, soft-hairy, viscid: leaves broadly lanceolate, coarsely serrate above: flowers glomerate at the ends of the branches: corolla bright salmon-color to orange, 18 lines long, tubular-funnel-form, segments acute. High valleys Fresno Co. Mrs. L. A. Peckenpah (nee Rawson).

HYDROPHYLLACEÆ.

Inflorescence usually scorpioid; flowers perfect, regular, 5-androus, the two styles distinct at least at the apex, except in *Romanzoffia* which has the stigmas as well as the styles united. Ovary commonly hispid or hirsute, at least at the top. Mostly herbs, with alternate or rarely opposite leaves and no stipules. In one of our genera the plants are shrubs, and in another they are more or less woody at the base. Key to genera and species p. 149.

ELLISIA.

Ellisia membranacea is in *Ray-Reg. Bot.* Nemophila membranacea; and E. chrysanthemifolia is Eucrypta chrysanthemifolia.

NEMOPHILA.

N. pedunculata, Benth. Only 2–4 inches high: corolla 2 lines or less in width, white with purple veinlets.

N. racemosa, Nutt. More slender than **N. aurita**, leaves shorter, nearly ovate in outline with fewer divisions and petiole not winged or with clasping base: flowers half as large.—San Diego.

PHACELIA.

§ 1. *Euphacelia*.

P. Circinata is **P. Californica** in *Ray-Reg. Bot.* Probably the correct name.

P. imbricata, Greene. (Next to **P. circinata**). Densely leafy at base: panicle of racemes in pairs, long and lax, the branches widely spreading: fruiting calyces compressed and closely imbricated; sepals unequal, the outer and larger deltoid-ovate, the others ovate oblong and ciliate.

P. nemoralis, Greene. Stout, erect, rather widely branching, 2–4 feet high, rough with stinging hairs, leaves simple or with a pair (rarely 2 pairs) of small leaflets at the base: pods 2-seeded.

P. sauveolans, Greene. Annual, branching from the base: soft, hairy and viscid, sweet scented: stem leaves oval, coarsely-toothed an inch long on slender petioles of nearly equal length, the lower leaves with some lyrate lobes at the base of the blade, corolla bright blue funnel-form, 6 lines long, 3 lines broad: sepals spatulate, 3 lines long.—Sonoma Co.

P. Eisenii, Brandegee. Annual, short, hairy, branching from the base 3–5 inches high: leaves 6–10 lines long on petioles as long, elliptic-oblong, simple or with a few basal lobes; flowers on slender pedicels twice their length: corolla about three lines long: stamens included: styles distinct.—Fresno Co. *Dr. Gustav Eisen.*

P. virgata, Greene. Stout, strictly erect. 2–3 feet high, leafy at base and to the middle, thence virgate-racemose: stem covered with a dense plushy coat of short hairs, with a sparse growth of bristly hairs: leaves pinnately divided into 2–3 pairs of lobes, a third the length of the terminal elliptic-lanceolate segment, appressed hairy: corolla small, dull yellowish: pod 2-seeded.

P. leptostachya, Greene. Annual, stout, widely branching, the branches often 2 feet long, roughish with a sparse growth of brownish hairs, slightly viscid: leaves ample, the lowest tripinnatifid: spikes usually solitary at intervals throughout the plant, in fruit 5–6 inches long: sepals spatulate, one much longer and twice as wide at the tip as the others: corolla small, little surpassing the calyx, dingy greenish white: stamens much exserted.—This species has heretofore been called **P. distans** or **P. tanacetifolia.** Prof. Greene in *Erythea.* Vol. II. p. 191, thus distinguishes these two species: "**True P. distans** is one of the commonest and most widely dispersed of Californian Phacelias. Its stem is more densely and quite retrorsely hispid. Its spikes are short and collected at or near the ends of the many branches in pairs or several together. Its corolla is

very broad and open, and of a lavender color. Its calyx is less unequal than that of
P. leptostachya." (In P. tanacetifolia the sepals are equal.)

§ 4. Eutoca.

P. *verna*, Howell. Annual, softly ashy with soft hairs, 4–10 inches high, branching
from the base and decumbent: leaves obovate to spatulate, entire or rarely the lower
most incisely toothed, abruptly contracted into a winged petiole or the upper sessile:
corolla pale blue but little exceeding the calyx: calyx lobes linear-lanecolate, 2–3 lines
long: seeds 8 to 12. Umpqua Val., Or.

P. leucantha, Lemmon. Annual, viscid, 1–2 feet high: leaves lanceolate, pinnatifid,
the segments entire or coarsely toothed: racemes ternate, dense: corolla limb rotate
nearly an inch broad pure white, the short throat and tube yellow: stamens very short:
seeds 20 to 25. Del Mar, San Diego Co., and with smaller flowers in San Luis Obispo Co.

§ 5. Microgenetes.

P. Cooperæ, Gray. Branched from the base 5–15 inches high, densely hairy viscid:
leaves oblong obtuse, crenately sub-pinnatifid, petiolate: corolla narrow, funnelform 2-
3 times longer than the calyx, the limb pink or violet, the throat and tube yellow: ovules
7 or 8.—Flood beds of streams, Santa Barbara and Ventura Co's.

Under § 3 of the Key to Genera in this order belongs,

LEMMONIA, Gray.

L. Californica, Gr. A depressed annual, hairy: stem branched from the base
dichotomously: leaves rosulate at the base and crowded at the ends of the branches,
entire, spatulate, 3–5 lines long: flowers sessile, solitary in the forks, closely cymose at
the ends of branches: calyx white-hairy: corolla white or whitish, a line long: styles
distinct. Mojave River. Recently found on Uncle Sam Mt., Lake Co., by Jepson.

ERIODICTYON.

E. Californicum, Greene, is E. glutinosum of this book.

NAMA.

N. stenocarpum, Gray. Annual (our other species perennial), hairy and slightly
viscid: diffusely branched, at length procumbent: leaves oblong or narrower, sessile,
entire. Los Angeles.

N. Lobbi, Gr. Woolly-hairy, resinous-viscous. This alpine species appears to
be near Eriodictyon, and is placed in that genus by Greene.

BORRAGINACEÆ.

Mostly roughly pubescent herbs, with alternate entire leaves without stipules, scorpioid inflorescence, and perfectly regular 5-androus flowers; the ovary of 4 lobes or divisions around a central style, ripening into seed-like nutlets. Calyx free, 5-parted or 5-cleft, persistent. Corolla with a 5-lobed limb, commonly imbricated in the bud. Stamens distinct, inserted in the tube or throat of the corolla alternate with its lobes. The one-sided and coiled apparent spikes or racemes straighten as the blossoms develop. Key to genera and species, p. 152.

Borrago officinalis has escaped from gardens in Santa Cruz. It is a very rough herb with clusters of nodding deep blue flowers, the rotate corollas and connivent anthers reminding one of potato blossoms.

LITHOSPERMUM.

L. arvense, Linn. Annual, a foot high, hoary with appressed hairs: leaves narrowly lanceolate or linear: flowers small white, sessile in leafy terminal cymes: nutlets conical, wrinkled. An Old World weed now apparently established in San Francisco.

Amsinckia lycopsoides. *a.* Calyx spread apart to show the ripe akenes.

AMSINCKIA.

A. campestris, Greene. Rather stout, 1-2 feet high, the short and rather dense spikes aggregated at the top of the stem: leaves linear-oblanceolate: sepals hardly twice the length of the nutlets: corolla inconspicuous: nutlets very dark brown, irregularly transverse-rugose and echinate-muricate. Byron Springs.

A. echinata, Gray. Erect, 1-2 feet high, very hispid with white spreading bristles: leaves linear lanceolate: sepals narrow, yellow-hispid: corolla small and very slender: nutlets almost prickly-muricate, not rugose. Perhaps not within our limits.

A. collina, Greene. Near **A. tessellata,** but slender and not branched: leaves narrowly linear-lanceolate, acute: calyx intensely gray-brown: corolla without folds in the throat: nutlets marked with few and sharp transverse ridges and intervening low tessellated granulations. Hills east of Livermore.

A. grandiflora, Kleeberger. Simple up to the terminal spikes, hispid: lower leaves oblanceolate, the upper lanceolate, all acute: sepals broad, often 4 or 3 only, tawny-hairy: corolla an inch long, deep yellow; the funnelform throat 6 lines long above the short proper tube which bears the nearly sessile anthers: nutlets light gray, sharply triangular, perfectly smooth and shining, the back straight or even concave. Antioch, hills east of Livermore, and north of Tulare Lake. This is **A. vernicosa** var. **grandiflora** of this book, but is undoubtedly a distinct species, and may well be called *Klee-berger's Amsinckia*, since (while a student in Yale College) he was the first to describe it.

KRYNITZKIA.

Sections 1 and 2 of this genus (species 1 to 6) and **Echinospermum Greenei** form the genus **Allocarya**, in *Bay Reg. Bot.* The following new species may be added to § 2.

K. stricta, Greene. Slender, erect and somewhat succulent: stem simple, or nearly so, 6 inches or less high, glabrous or nearly so, all except the floral leaves opposite: flowers very small: calyx lobes closed over the growing fruit: nutlets light gray, shining, numerous close transverse ridges. Calistoga.

K. diffusa, Greene. Procumbent, a foot or less long, loosely branching from the base; racemes leafy for half their length; lowest pedicel 6 lines long, the others hardly a line: calyx widely spreading: flowers small, nutlets dark brown, broadly ovoid, incurved, the back with rather sharp granulations and ridges.

K. humistrata, Greene. Stout and succulent, the branches mostly prostrate, a foot long: pedicels short and thick, often diflexed: calyx lobes in fruit becoming 4–6 lines long, turned to one side so as to stand vertically in a row: nutlets with minute muriculations and sharp transverse wrinkles which have tufts of minute bristles. This is *K. Californica*, var. *subglochidiata.* Probably a good species.

K. myriantha, Greene. A diffuse, slender, prostrate or trailing annual a foot or more long: lower flowers on short, slender pedicels, the others forming close spikes: nutlets distinguished from those of *K. Chorisiana* by narrower outline, greater length, a more glossy surface and more prominent ridges on the back. Perhaps the more floriferous form of the latter species. Monterey.

K. vestita, Greene. Distinguished from *K. mollis* by stout, nearly erect annual stems 2 feet or more high, rather rough pubescence and dark nutlets reticulated, the scar surrounded by a ridge. Petaluma.

K. plebeia, Gr. Depressed branches a span or more long: floral leaves linear-oblong: nutlets ovoid a line long, the back rugose-reticulate glabrous, not granulate or muricutate. Humboldt Bay, C. C. Marshall.

K. Austinæ, Greene. Erect, slender, simple or a few branches, about a span high, nearly glabrous, except the calyx: leaves narrowly linear 1½–3 inches long: nutlet ovate-acuminate, strongly keeled on both sides, the dorsal keel and margins with stout prickles, the uppermost barbed. Butte Co.

K. stipitata, Greene, Ten to 18 inches high, erect and simple or with ascending branches from the base, light green, nearly glabrous: calyx nearly sessile, segments spreading foliaceous, in fruit often 6 lines long: corolla short-funnelform, 3–6 lines broad: nutlets slender-ovoid, the back covered with blunt tubercles and transverse wrinkles, divergent, stipitate. This, according to Greene, is one of the most common species in Central California. Moist land.

K. Hickmanii, Greene. Very slender, diffuse, the filiform racemose branches 6–10 inches long: calyx a line long, on filiform pedicels: corolla a line or more broad: nutlets dark colored, tuberculate. Monterey Co.

K. hirta, Greene. Annual, more than a foot high, erect, flaccid, simple below with many pairs of connate-sheathing linear leaves, loosely racemose above, bristly hairy: racemes in pairs: pedicels slender, a line long: calyx lobes erect, very hairy: corolla 3 lines broad: ovoid nutlets dark colored, the back granulate and obscurely wrinkled. Umpqua Valley, Or.

K. scripta, Greene. Somewhat succulent strigose-hairy: branches prostrate, 6–10 inches long: pedicels stout in the axils of leafy bracts: sepals oblanceolate at length, standing vertically in row: nutlets a line long, the back dark and smooth, marked by a sharp irregular flexuose with white ridges, these beset with tufts of short spreading bristles.

Section 3 is equivalent to **Cryptanthe** in *Bay-Reg. Bot.* The following species are to be added:

C. flaccida, Greene, is **K. oxycarya** of this book.

K. Clevelandi, Greene. A foot or more high with few ascending branches rough with bristly hairs: calyx slender, appressed to the rachis: nutlets 2 or 1, shining.

K. hispidissima, Greene. Size of the last, but more branching, more densely hispid with softer hairs, and the inflorescence more elongated: corolla conspicuous: nutlets 4, similar to those of *K. leiocarya*, much surpassed by the slender calyx lobes. San Luis Obispo Co.

K. nemaclada, Greene. Slender, very diffusely branching, a foot high, sparsely bristly-hairy: spikes very loose, almost filiform: calyx a line long appressed to the rachis, bristly: nutlets 4, ovoid-acuminate $\frac{1}{2}$ a line long, shining. Colusa Co.

K. Rattani, Greene. Hispid with slender hairs and slightly canescent, about a foot high, slender but rigid: leaves linear: spikes in 3s on an elongated naked common peduncle, rather densely flowered: calyx appressed to the rachis, its bristles spreading and straight: nutlets (3 maturing) lance-ovoid $1\frac{1}{2}$ lines long, brownish and smooth but not shining. First collected by Hickman in Monterey Co.

K. crinita, Greene. Annual, slender 8–12 inches high, somewhat fastigiately branched from the base, rather stiffly hirsute: leaves linear: dense spikes elongated:

calyx about 4 lines long, densely white-hairy; nutlet solitary ovoid, the dull brown surface smooth but not polished. Shasta Co.

SOLANACEÆ.

Herbs or shrubs, with alternate leaves and no stipules, regular 5-merous flowers on bractless pedicels, a single style and a 2-celled ovary; the fruit a many-seeded berry or capsule. Key to genera and species, p. 157.

This small order of, perhaps, not more than twenty species west of the Sierra Nevada, and less than 70 in North America, is remarkable for the diversity of properties exhibited by its members, and the almost universal use by man of several of its species. At first view, the classification seems absurd which puts fiery cayenne pepper and insipid egg plants, the wholesome tomato and deadly nightshade, nutritious potatoes and poisonous tobacco together in one family. A careful examination shows that these seemingly very different plants are much alike after all. The four most important plants of this order—potato, tobacco, red or Cayenne pepper, and tomato—are natives of tropical America, and were consequently not used in the Old World before the sixteenth century. The following ornamental plants of the order are common in cultivation: Jerusaleum Cherry (*Solanum Pseudo-Capsicum*, a small shrub with red berries; Jasmine Solanum (*S. jasminoides*), a shrubby climber, with a profusion of nearly white blossoms a little smaller than those of the potato; the well-known Matrimony Vine (*Lycium vulgare*); Tree Datura or Stramonium (*Datura arborea*), with hanging flowers six or seven inches in length; *Cestrum*, a shrub with drooping tubular red flowers in terminal bunches; and *Petunia*, with funnel-form corollas of various colors.

SOLANUM.

S. elæagnifolium, Sav. A low perennial, silvery, whitened by a dense coat of stellate hairs, often with small prickles: calyx 5-angled, lobes slender: corolla violet, an inch or less broad. Tulare Co.

S. villosum, Lam. Annual, hairy: leaves an inch long or more, sinuate-dentate: corolla white, minute. Introduced.

S. alatum, Moench. Similar but with angular stem and red berries. Introduced.

S. cupuliferum, Greene. Distinguished from **S. umbelliferum** by leaves transversely rugose, margin crisped, hairs with pustulate base and flat corolla.

SCROPHULARIACEÆ.

A corolla more or less bilabiate, with the lobes imbricated in the bud; didynamous or diandrous stamens; a single style and a 2-celled ovary and capsule mark this large order. In *Pentstemon* there is a fifth rudimentary stamen. *Verbascum* has five perfect stamens. Key to genera and species, p. 158.

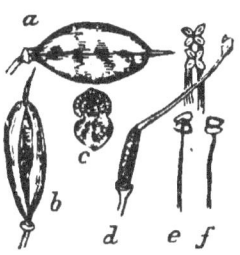

Mimulus glutinosus.

a. Ripe capsule of Mi-
mulus luteus. b. The
same seen edgwise burst-
ing open. c. A cross
section of the same, showing the placentæ and
seeds. d. Pistil of Mi-
mulus luteus. e. Front
view of one of the an-
thers. f. Back view of
the same. Above these are the stamens of Mimulus glutinosus united in pairs.

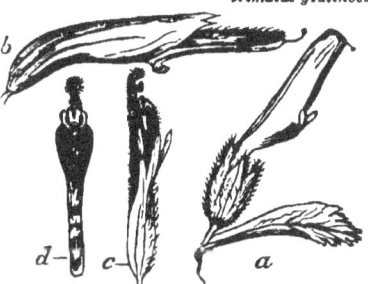

a. Single flower and bract of Pedicularis densiflora (galea flattened laterally, the pistil protrud-
ing; the lower lip of 3 small lobes, 2 of which
are shown.) b. A single flower of Castilleia.
c. Single flower of Orthocarpus purpurascens.
d. Front view of the same, with calyx removed.
The lower lip (anterior or front part of the flower)
3-lobed, the galea beaked and surpassing the
stigma.

This large order, numbering nearly 2,000 species, is remarkable for the great beauty
of its flowers, and for the impartial distribution of its species over the whole world.
Over 375 species, belonging to 37 genera, are natives of the United States. About 75
species grow east of the Mississippi, and about 175 west of the Sierra Nevada in this
State.

LINARIA.

L. vulgaris, Mill. Is occasionally found by roadsides. Stems very leafy: flowers
yellow, an inch or more long, in a dense raceme. A native of Europe, often called
Butter and eggs.

COLLINSIA.

C. concolor, Greene.   Near **C. bicolor** and probably only a variety distinguished by fewer red-purple flowers half as large, hairy calyx and more slender, revolute leaves. San Diego Co.

C. Franciscana, Bioletti.   Is a form described as intermediate between **C. bicolor** and **C. sparsiflora**: slender with thinnish foliage: leaves ovate or narrowed: pedicels 1-6 in the axils of upper leaves, from shorter, to 2 or 3 times as long as the calyx: corolla 9 lines long, bluish dotted with purple, throat closed at the mouth.   San Francisco Bay region.

C. arvensis, Greene.   Erect, simple or with several nearly erect branches from the base, 10-18 inches high, glabrous except the very sparsely bristly-hairy leaf-margins: lowest leaves oval or oblong on petioles equaling the blade, 6 lines long, coarsely toothed or nearly lobed; stem leaves lanceolate to linear, sessile, revolute: flowers loosely racemose, deep violet, 9 lines long: calyx-teeth lanceolate-subulate, twice the length of the tube: corolla with compressed throat 3 lines long and broad, upper lip half the length of the lower and paler: filaments slightly hairy below.   This has usually been referred to *C. sparsiflora*, which species Greene restricts to a small flowered form.

C. Wrightii, Watson.   Divaricately branched, 3 or 4 inches high, glandular-hairy, viscid: leaves linear-oblong to lanceolate, entire: flowers pedicellate, 1 or 2 in the axils, the uppermost in a naked umbel: calyx 2 lines long: corolla but a little longer, throat and tube shorter than the broad lobes; lower lip blue, upper yellowish: pod 2-seeded. Near *C. Torreyi*.   Alpine, in Kern Co.

TONELLA.

In *Bay-Reg. Bot.* this genus is united with **Collinsia**, and our species is there called **C. tenella**, Benth.

PENTSTEMON.

P. leucanthus, Greene.   Erect, 4-6 feet high, pallid, glaucous: leaves linear-lanceolate, entire: sepals ovate with acuminate tips: corolla white, 12-18 lines long, limb with short spreading lobes: anthers horseshoe-shape, filaments naked.   Santa Barbara Co., John Spence.

P. Sonomensis, Greene.   Suffrutescent, evergreen, very leafy, 5-10 inches high, slightly puberulent: leaves rather light green, coriaceous, denticulate, 6-9 lines long, short-petiolate, the uppermost round-ovate acutish, lower nearly orbicular and retuse: corolla 1 inch long, deep red, segments nearly equal, not widely spreading: anther slightly exserted, white, wooly.   Better considered a variety of **P. Menziesii**, Hook., of which the type belongs on the North Coast.

MIMULUS.

M. glutinosus is **Diplacus glutinosus** in *Bay-Reg. Bot* ₃ 2, § 3, and § 4 make the genus **Eunanus** in *Bay-Reg. Bot.*, and **M. exilis** is **Mimetanthe pilosa**, Greene.

§ 3. Eunanus.

Mimulus Austinæ, Greene. Near **M. mephiticus**, but odorless and scarcely viscid, 1-3 inches high, much branched, leaves spatulate 6 lines long or more, entire with 3-5 parallel veins: calyx teeth nearly equal: corolla yellow, throat purple dotted, tube long and slender, limb broad: pod attenuate, greatly surpassing the calyx. Modoc Co.

M. Cusickii, Greene. A foot high or less, simple or branched; leaves broadly-ovate, very acute, entire sessile, 1 inch long or longer: calyx teeth very unequal, acute: corolla red-purple, tube slender, limb rotate, nearly regular, 6-10 lines broad. Or. and Wash.

§ 5. Eumimulus.

According to Greene **Mimulus luteus** is a South American species quite distinct from our species which has borne this name, and should be known by the name given to it long ago by an eminent French botanist; viz.

M. guttatus, D. C. A very variable species, of which two varieties described by Greene are the most common; var. **grandis**, Greene, a stout perennial 2-5 feet high with round stems, usually simple above the decumbent base; the stem leaves orbicular to round-ovate, and those from the base petiolate and sometimes lyrate; the flowers an inch or more long, light yellow with red dotted throat, in racemes a foot or more in length: var. **insignis**, Greene, probably always annual, less than 2 feet high, the flowers with red dotted calyx and corolla with large, dark red spots. But there are too many intermediate forms to make these names worth much.

M. arvensis, Greene. Annual, erect, simple, quadrangular stems, 1-2 feet high or more: lower leaves coarsely toothed and pastate or lyrate, the floral soft-hairy beneath: calyx 3-4 lines long, purple dotted, nearly truncate, becoming in fruit 6-8 lines long: pod compressed, nearly orbicular.

M. subreniformis, Greene. Slender, 2-6 inches high: leaves 2-5 lines long, reni. form, with remote teeth, purplish beneath, roughish above, with short white hairs: corolla little exceeding the calyx, yellow with dark dots. Shasta Co.

M. marmoratus, Greene. Decumbent, annual, 4-8 inches long, sparsely hairy: stem acutely angled: leaves ovate, reniform, toothed, red beneath, 6-9 lines long: calyx mottled, 4-5 lines long: corolla an inch or less in length, with slender tube and ample limb, the middle lobe large, hairy, with a red spot and dots. Knight's Ferry. F. W. Bancroft.

M. deflexus, Wat. Slender, 2-3 inches high: leaves linear or broader, entire, less than 6 lines long: pedicels longer, spreading or reflexed: calyx slightly angled, nearly equally toothed, less than 3 lines long: corolla 6-8 lines long, upper lip deep purple, lower deep yellow, with bifid lobes, somewhat hairy and spotted below. Near **M-bicolor**. Mts. of Tulare.

M. latidens, Greene. Annual, slender, much branched from the base, 3–10 inches high, glandular, soft-downy, leaves ovate or narrower, 6–12 lines long, 3–5 nerved, entire or sparingly toothed: calyx 3–6 lines long: corolla 3–5 lines long, nearly regular but limb very small, white.

§ **6. Mimuloides.**

Greene, in *Bay-Reg. Bot.*, makes a genus of this section, **M. exilis** being there called **Mimetanthe pilosa**, Greene.

CASTILLEIA.

C. parviflora, Bong., is **C. Douglasii**, Benth., in *Bay-Reg. Bot.*

ORTHOCARPUS.

O. versicolor, Greene. Slender, slightly reddish, mostly simple, 6 inches high or less: corolla with a shorter tube, and the sacs much larger than in **O. erianthus**, white fading pinkish, throat densely bearded. This was considered a form of **O. erianthus** by Gray. Greene puts the var. **roseus** here.

Plantaginella, in *Bay-Reg. Bot.*, is **Limosella** in this book, **Wulfenia** is **Synthyris**, and **Adenostegia** is **Cordylanthus** (the species **rigida** is **C. filifolius**).

BELLARDIA, Allioni.

B. Trixago, All. Stout, rigid, erect, 1 foot high or more: leaves lanceolate, crenate-serrate: spike thick, dense, 4-sided, several inches long: corolla 6–12 lines long, rose color and white, strongly bilabiate: upper lip enclosing in its concavity the 4 stamens surpassed by the 3-lobed lower lip. Escape from gardens near Martinez.

CONVOLVULACEÆ.

Herbs, usually twining or trailing, with alternate leaves (or scales) and regular perfect flowers; the stamens as many as the lobes or angles of the corolla and alternate with them (5, rarely 4); the free persistent calyx of mostly distinct imbricated sepals; ovary 2–3 celled; capsules generally globular; seeds 1 to 6. Inflorescence axillary. Key to genera and species, p. 156.

DICHONDRA.

D. repens, Forster. Prostrate: leaves round-reniform: small flowers axillary; corolla deeply 5-lobed, yellow: carpels 2, distinct, hairy. Presidio, San Francisco.

CONVOLVULUS.

C. Binghamæ, Greene. Perennial from creeping root stocks, the stems 3–6 feet long: leaves glabrous, oval or oblong acute, the base with a pair of obtuse hastate lobes: bracts oval to narrowly oblong, 4 lines long closely embracing the calyx which is twice as long· corolla white. In marshy places, Santa Barbara. Mrs. R. F. Bingham.

C. occidentalis. According to Greene this species is distinguished by broad keeled bracts which completely cover the calyx, and flowers a third larger than those of **C. luteolus,** which has small bracts growing 6 lines or more below the flower. The former grows south of Monterey, and the latter north. But Gray considered the latter as a form of **C. occidentalis.**

C. subacaulis of *Bay-Reg. Bot.* is **C. Californicus,** Choisy.

OROBANCHACEÆ.

Root-parasitic herbs, destitute of leaves and green color. Distinguished from *Scrophulariaceæ* by the 1-celled ovary: the placentæ parietal. Key to genera and species, p. 169.

LENTIBULARIACEÆ.

Aquatic herbs. Upper lip of corolla interior in the bud: lower lip with a palate projecting into the throat and a spur, 3-lobed. Key to genera and species, p. 169.

RIGHT-HAND FIG.—*a.* Flower of Sphacele calycina. *b.* Same with corolla cut to show stamens, pistil, and hairy ring inside. *e.* Fruit (4 globular akenes) of the same, lying in the bottom of the calyx. *c.* Flower and buds, showing inflorescence of Trichostema lanceolatum. *d.* Ripe fruit of the same (4 triangular akenes in the persistent calyx).

LABIATÆ.

Chiefly aromatic herbs with square stems, opposite simple leaves, and no stipules, bilabiate corolla, didynamous or diandrous stamens, and a 4-lobed ovary with a single style, forming seed-like nutlets in the bottom of the persistent calyx. Flowers perfect, axillary. Calyx 3-5 toothed or cleft, or bilabiate. Stamens on the tubes of the corolla. Style, 2-cleft at the apex, often unequally so, or one of the lobes obsolete: stigmas minute. Key to genera and species, p. 170.

MENTHA.

The following mints have run wild in moist places:

M. viridis, L. (Spearmint.) Green and glabrous, erect: leaves oblong-lanceolate, acute, incisely serrate, or nearly so: flowers crowded in terminal spikes.

M. piperita, Huds. (Peppermint.) Darker green: leaves petiolate, ovate to lanceolate: flowers in shorter, thicker spikes.

M. Pulegium, L. (Pennyroyal.) Prostrate, much-branched: leaves smaller than those of the preceding, usually less than an inch long: calyx with a hairy ring.

MONARDELLA.

M. Sheltonii, Torr, in *Bay-Reg. Bot.* is a variety of *M. villosa,* Benth.

PYCNANTHEMUM.

P. Californicum, Torr, is **Kœllia Californica** in *Bay-Reg. Bot.* Whitish with soft hairs: leaves ovate or narrower, sessile: flowers white.

MICROMERIA.

M. purpurea, Gr. Greene thinks this is the common garden *pennyroyal* (**Mentha pulegium,** L.), but the description of the former in the Synoptical Flora of North America does not fit the latter plant.

MELISSA, Tourn.

M. officinalis, L., Common Balm, may be found outside of gardens occasionally. Its lemon-like odor distinguishes it from other mints.

AUDIBERTIA and SALVIA.

These genera are joined under the latter name in *Bay-Reg. Bot.,* and a species of the former (**A. polystachia**) is the basis of a new genus **Ramona,** Greene, to which is also added **A. humilis,** which is the species described in *Bay-Reg. Bot.*

NEPETA, Linn.

N. cataria, L. (Catnip) and **N. Glechoma,** Benth, (Ground Ivy) are found occasionally. The former is erect, 2 feet high or more, somewhat hoary with minute downy hairs; the leave petiolate, ovate-cordate, coarsely toothed; the small flowers in short. dense spikes: calyx downy, 1-5 ribbed, teeth nearly equal; corolla bluish or nearly white; the upper pair of stamens projecting above the lower pair. The latter is a creeping plant with orbicular-reniform, crenate leaves; the blue flowers 9 lines long, usually six in a whorl in the axils of leaves on short, ascending branches.

LAMIUM, Tourn.

L. amplexicaule, L., is another European herb fairly established in San Jose and some other localities. Stems weak, leaves distant, the lower petioled, the upper sessile or clasping, crenately lobed and incised: slender purple corolla with spotted lower lip: anthers hairy.

STACHYS.

S. velutina, Greene. Stout, 2–6 feet high, soft-hairy: leaves rugose, short-petioled, cordate-ovate, 3–6 inches long: calyx-teeth spreading, corolla small, white with red dots and lines; upper lip deeply concave concealing the stamens. Too near **S. albens.**

S. stricta, Greene. More slender than the last and smaller, less hairy, resinous-glandular: leaves narrower and thinner: corolla white, the upper lip sub-orbicular, slightly concave, and not concealing the stamens, the back hairy; the lower lip with side lobes mere deflexed teeth.

S. Californica, Bentham. Distinguished from **S. bullata** by larger size (3–6 feet high), aromatic odor, ovate-cordate leaves and corolla darker. Perhaps only a moist ground variety of the latter.

Verbena hastata (spikes and bracts). *a.* Ripe fruit removed from the calyx (natural size and magnified).

VERBENACEÆ.

Herbs or shrubs differing from *Labiatæ* mainly in the ovary and fruit, which is undivided and 2–4-celled, at maturity either dry and splitting into as many 1-seeded nutlets, or drupaceous, containing as many little stones. Key to genera and species, p. 170.

LIPPIA.

L. cuneifolia, Steud. Diffusely branched from a woody base, procumbent, canescent: leaves rigid, cuneate-linear sessile, incisely toothed above the middle: peduncles short, bearing cylindrical heads 4 or 5 lines thick. River banks and subsaline plains of the Central Val., *Greene.*

PLANTAGINACEÆ.

This order is represented in North America by

1. PLANTAGO, L. PLANTAIN.

Flowers in spikes or heads, bracteate. Calyx of 4 persistent sepals free from the ovary. Corolla scarious, apparently dry, colorless, 2-lobed. Stamens 2 or 4 on the corolla alternate with its lobes, anthers versatile. Style filiform, bearded above. Stemless herbs with nerved or ribbed radical leaves and naked scapes of small greenish or colorless flowers. Key to genera and species, p. 174.

DIVISION 3. APETALÆ.

ARISTOLOCHIACEÆ.

Twining shrubs or low herbs with perfect flowers, the conspicuous lurid calyx valvate in the bud and coherent with the 6-celled ovary, which forms a many-seeded 6-celled pod or berry in fruit. Stamens 6–12, more or less united with the style; anthers adnate, extrorse. Leaves petioled, mostly heart-shaped and entire. *Gray's Manual.*

ARISTOLOCHIA.

Calyx tubular, inflated above the ovary. Stamens 6, the sessile anthers adnate to the short stigma.

A. **Californica**, Gr. (Pipe-Vine.) A twining shrub with large cordate leaves, flowers curved like a Dutch pipe, greenish, marked with brown or purple.

2. ASARUM. Tourn.

Calyx regular, 3-cleft or parted. Stamens 12, with more or less distinct filaments their tips usually continued beyond the anther into a point. Stemless herbs with creeping rootstocks, bearing 2 or 3 scales, then one or two leaves, and terminated by a short peduncled flower close to the ground.

A. **caudatum**, Lindl. (Wild Ginger.) The smooth, broadly cordate leaves usually mottled with white; calyx bell-shaped, the acuminate lobes spreading, brownish purple. Common in forests; the flowers likely to be hidden under leaves.

A. **Hartwegi**, Watson. Stouter than the last: leaves cordate, mottled margin ciliate: peduncle stout, 6 lines long: calyx-lobes ovate, narrowed to a linear apex 1–1½ inches long: anther shorter than the appendage or produced filament above.

A. **Lemmoni**, Watson. Leaves rounded at apex, flat: calyx lobes 4–6 lines long. Plumas Co.

NYCTAGINACEÆ.

Herbs with mostly opposite and entire leaves, stems swollen at the joints, the tubular calyx corolla-like, its persistent base contracted, inclosing the 1-celled, 1-seeded ovary, and becoming a sort of indehiscent pod. Flowers with an involucre encasing from one to many flowers.

MIRABALIS, Linn.

Involucre calyx-like, 5-cleft, 1-12-flowered: perianth (corolla-like calyx) tubular or funnelform with a spreading limb: stigma capitate: ovary globose to oblong, smooth or ribbed. Perennial herbs with leaves nearly equal in the pairs.

M. Frœbelii (Behr), Greene. Stout, spreading stems 2-3 feet long, viscid-hairy: leaves broadly ovate or narrower, the lower cordate often 4 inches long: involucre an inch long, usually 6-flowered: perianth broadly funnelform, about 1½ inches long, purple. Southern Cal.

M. Greenei, Watson. Similar to the last with thicker, larger leaves, the involucre 7-10 flowered. Northern Cal.

M. lævis (Benth), Curran. Viscid-hairy, yellowish green: leaves rounded ovate to cordate, 6-15 lines long: involucre 2-3 lines long; perianth 5 lines long. Southern Cal.

M. Jalapa is the cultivated *Four-o'clock* or *Marvel-of-Peru.*

ALLIONIA, L.

A. incarnata, L. Slender, prostrate, woolly-hairy, viscid: leaves unequal pairs: involucre 3-flowered, 2-3 lines long: perianth 2-4 lines long, one lobe much shorter than the others, purple or white. Monterey S.

ABRONIA, Juss.

Calyx salverform, with obcordate lobes. Stamens 5, included, adnate to the tube. Style included; stigma, capitate or clavate. Fruit 2-5-winged. Embryo by abortion monocotyledonous, enfolding mealy albumen. Low herbs, with the opposite thick petioled leaves unequal, and the flowers in involucrate heads. Common on sandy sea beaches. A viscid exudation causes sand to stick to every part of the plants.

A. latifolia, Esch. (Yellow Sand-Verbena.) Root perennial; stems procumbent; leaves very thick, sub-cordate to reniform, on thick petioles; flowers orange-yellow, fragrant.

A. umbellata, Lamb. (Pink Sand-Verbena.) Annual; stems decumbent, leaves oblong or ovate, attenuate at base into slender petioles; flowers pink.

A. maritima, Nutt. (Red Sand-Verbena.) Stouter than the last; leaves broader with shorter petioles; involucral bracts ovate; flowers bright red. From Santa Barbara southward.

A. fragrans, Nutt., of the Columbia River, has white flowers.

POLYGONACEÆ.

Herbs, with alternate entire leaves, and stipules in the form of sheaths, or obsolete, above the swollen joints of the stem; the flowers mostly perfect, with a more or less persistent calyx, a 1-celled ovary, bearing 2 or 4 styles or stigmas, and a single seed. Stamens 4–12 inserted on the base of the 3–6-cleft calyx. Rhubarb and buckwheat are the only wellknown plants of this order in cultivation.

POLYGONUM.

Calyx 5 parted; the divisions petal-like, persistent in fruit, and surrounding the usually 3-angled akene. Stamens 3 to 8. Styles or stigmas 2 or 3. Herbs with small flowers on jointed pedicels.

Knot-weed or Yard-grass and Smart-weed belong to this genus. About 25 species are found in California, of which 2 or 3 are introduced weeds. Two or three species are useful forage plants (Saccaline, Knotgrass.)

RUMEX, L.

Calyx of 6 sepals; the three outer herbaceous, spreading in fruit; the three inner larger, somewhat petaloid, covering the akene in fruit (then called valves), and often bearing grainlike appendages on the outside. Stamens 6. Styles 3; stigmas tufted. Introduced weeds with small greenish flowers crowded and whorled in panicled racemes.

The Docks and Sheep-sorrel are examples of this genus. Of the dozen species on this coast, half are introduced weeds. Canaigre, cultivated for the use of tanners, is a kind of Dock, native of this coast (R. hymenosepalus).

ERIOGONUM, Michx.

Flowers borne in a many-to-few-flowered calyx-like involucre of united bracts: the pedicels exserted, jointed to the flower, with bractlets at the base: calyx corolla-like, 6-parted or deeply 6-cleft: stamens 9: akene triangular. Herbaceous or somewhat woody plants, usually with a woolly or scurfy pubescence: the entire leaves without stipules and mostly radical: juice frequently acid. Over 80 species grow west of the Mississippi, of which over 50 are Californian, mostly Alpine.

CHORIZANTHE, R. Br.

Flowers borne in 1–3 flowered involucres, which have 3–6 awned segments or teeth, the tube ribbed or angled. Stamens 9 (rarely 6 or 3): akene triangular. Annuals branching dichotomously, the leaves few and mostly basal, ternate bracts at the joints. More than 20 species on this coast.

OXYTHECA, Nutt.

Flowers borne in few-flowered, pedicellate, awn-tipped or unarmed involucres, Perianth 6-parted: stamens 6: akene commonly lenticular. Slender annuals, branching dichotomously, the leaves in a rosulate tuft at the base, bracts united at the base and leaf-like. Half a dozen species on this coast.

PIPERACEÆ.

Herbs with jointed stems, alternate entire leaves and perfect flowers in spikes, entirely destitute of floral envelopes.

ANEMOPSIS, Hook.

Flowers in a simple conical spadix, which is surrounded by a 5–8-leaved persistent colored involucre, each flower subtended by a free colored bract. Stamens 6 to 8, free, growing upon the immersed ovary.

A. Californica, Hook. Stem simple, erect, 3 to 15 inches high, with a single broad, clasping leaf in the middle, and an axillary branchlet reduced to 1 or more petioled leaves; radical leaves oblong-oval, cordate at base, 2 to 6 inches long; involucre 1 to 1½ inches broad, white, becoming brown. Used medicinally by the Mexicans, who call it *Yerba Mansa.*

APETALOUS TREES.

The Order **Betulaceæ** (Birch Family) is represented in California by two Birches, which scarcely attain to the dignity of trees, and are confined to the high Sierras, and four Alders, two of which grow in the central part of the state, viz.:

Alnus rubra, Bong. (Red Alder), and the more common

Alnus rhombifolia, Nutt. (White Alder), which may be distinguished by its thinner leaves, not rusty beneath, and more slender branches not so distinctly dotted with white.

Myrica Californica, Cham. (Bayberry) representing the Order **Myricaceæ**, grows in moist places, and may be known by its thick oblanceolate serrate evergreen leaves and dense clusters of small fruit, whitened by a coat of wax.

Umbellularia Californica, Nutt. (Order **Lauraceæ**), is the well-known Laurel.

Platanus racemosa, Nutt, is the California Sycamore.

The Order **Salicaceæ** is represented by 4 or 5 Willows, large enough to be called trees, and 3 Poplars, viz.:

Populus tremuloides, Michx. (Quaking Asp), a small tree, with whitish bark and round ovate leaves. In the high Sierra. The only Californian tree, except one or two willows, found east of the Rocky Mountains.

P. trichacarpa. Torr & Gr. (Cottonwood.) Petioles round; young bark brownish.

P. Fremonti, Wat. (Fremont's cottonwood.) Petioles flattened; young bark yellowish.

The WALNUT FAMILY is represented by **Juglans Californica**, the California Black Walnut.

About a dozen kinds of Oak Trees, and several shrubs of the same genus, with the chestnut-like *Chinquapin*, represent the Order **Cupuliferæ**.

CLASS II—ENDOGENS OR MONOCOTYLEDONS.

Stems consisting of woody tissue and cellular tissue (pith) intermixed.   Embryo mono-cotyledonous.

ALISMACEÆ.

Marsh herbs, with leaves all radical, scape-like flowering stems, and (in our species) perfect flowers.   Sepals and petals each three and distinct.   Ovaries 3 to many; distinct, or, at least, separating at maturity, forming 1–2-seeded pods.   Stamens from 6 to many; anthers extrorse, 2-celled.   Key to genera and species, p. 174.

ORCHIDACEÆ.

Herbs with irregular 6-merous perianth adnate to the 1-celled ovary; the ovules innumerable on 3 parietal placentæ, becoming fine sawdust-like seeds.   One petal, called the lip, is unlike the other two.   Stamens consolidated with the style forming the *Column.* This remarkable family of plants is chiefly tropical, one only—**Calypso borealis**— reaches the limits of the Arctic Circle.   Most of the tropical species are epiphytes. These cling to other plants, usually trees, by means of aerial roots, which, however, take no nourishment from the supporting plants.   More than 2,500 kinds of epiphytal orchids are known, mostly South American.   These are often remarkable for the beauty as well as oddity of their flowers, characters which make them the most admired of hothouse plants.   But the wonderful mechanism of the flowers, by means of which insects effect cross-fertilization, is more interesting to the naturalist than perfume and beauty, which are the more common agents used by higher plants to ensure this aid of insects in the production of good seed.

The only plant production of this order well known in commerce is vanilla, the fleshy pods of several creeping or climbing species of the genus **Vanilla**, all natives of Mexico, Colombia and Guiana.   Key to genera and species. p. 175.

HABENARIA.

H. maritima, Greene, is more robust than *H. elegans,* with a short, thick spike of whiter, larger flowers; the lip pure white.

H. Michaeli, Greene.   Still more robust, the fleshy stem bearing many triangular or ovate acute, thin appressed bracts, the spike of greenish flowers 3 inches long; sepals and petals longer, 3 lines long: spur a third longer than the ovary.   These two species may be forms of *H. elegans.*

H. saccata, Greene.   Two or more feet high, with a slender, leafy bracted raceme of green flowers; the side petals falcate, the linear lip much larger than the saccate spur.   May be a form of **H. gracilis.**

SPIRANTHES is **ORCHIASTRUM** in *Bay. Reg. Bot.*, and **Epipactis** is Limodorum.

IRIDACEÆ.

Herbs with 2-ranked leaves, the flower buds inclosed by bracts. Perianth adherent to the ovary, segments in two, often unequal sets. Stamens 3. anthers extrorse. Ovary 3-celled, style 1, stigma 3, often petaloid. Key to genera and species, p. 177.

SISYRINCHIUM.

This genus is **Bermudiana** in *Bay-Reg. Bot.*
S. sarmentosum, Suksdorf. Stem and leaves very slender: segments of the small light blue perianth, all abruptly acuminate. Washington.

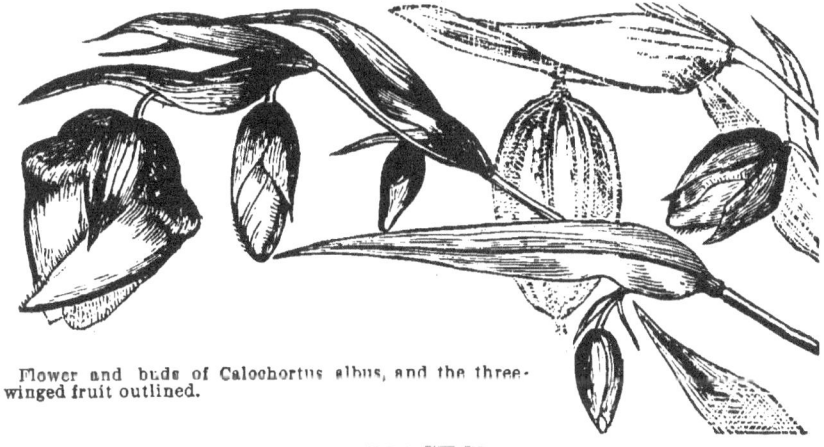

Flower and buds of Calochortus albus, and the three-winged fruit outlined.

LILIACEÆ.

Herbs, or rarely woody plants, with regular and symmetrical flowers; the perianth free from the chiefly 3-celled ovary, with the divisions all petaloid (except in *Trillium* and *Calochortus*), the stamens opposite the divisions of the perianth, usually 6 (rarely 3 of them, not pollenifrous), sometimes only 3, and in one genus the parts of the flower in 2's. The antherless stamens in about a dozen of our species are not rudimentary, as in **Collinsia**, but well developed, as in **Pentstemon**, and probably as in that genus, assist insects in cross-fertilization. These stamens are called *staminodia*. Key to genera and species, p. 178.

ALLIUM.

A. dichlamydeum, Green   Leaves few, shorter than the scape, which is about a foot high: perianth deep rose color 5 lines long, outer segment spreading, inner erect, entire: ovary crested.   Resembling *A. serratum,* which has smaller flowers with serrulate segments.

A. crispum, Greene.   Scape 5–8 inches high, thickened upward: spathe 1-valved pedicels 12–25, 1 inch long: perianth light purple, 3–4 lines long, the outer segments oblong-ovate plane, entire, the inner lanceolate, the margins undulate: ovary not crested. Near *A. serratum.*   Paso Robles.

A. monospermum, Jepson.   Size and habit of **A. attenuifolium,** but 3 or 4 scapes from the red bulb: pedicels 50–80: perianth pale purple: filaments with broadly deltoid united bases; pod maturing but one seed.   Vac. Mts.

BRODIÆA.

Prof. Green has again elaborated this genus and its allies, but we shall retain the names of his first revision.   In *Bay-Reg Bot.* **Dichelostemma** replaces **Brodiæa** and **Brevoortia** is added, becoming **D. Ida-Maia,** while **B. volubis** becomes **D. Californicum.** The other species retain the old specific names with the final "a" changed to "um," since **Dichelostemma** is neuter.

Brodiæa capitata.

HOOKERA.

H. leptandra, Greene.   Scape slender, a foot or less high: umbel 2-flowered: perianth purple, an inch long; segments linear, spreading above free portion of the filaments 3 lines long: anthers linear, 3 lines long; staminodia thin, involute, retuse.

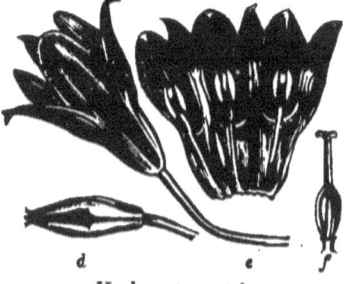

d   e   f

Hookera terrestris.

TRITELEIA.

The yellow flowered species of this genus appear in *Bay-Reg. Bot.* under the generic name **Caliprora,** and the white flowered one is **Hesperoscordum.**

T. Hendersoni, Watson.   Near **T. Bridgesii.** Scape and leaves about a foot long: leaves 3–5 lines broad: perianth salmon-color with brown-purple nerves, 6–9 lines long, the segments about

Triteleia Hya-cinthina.

Triteleia laxa.

as long as the tube: stamens in one row, the filaments equal. Oregon.

T. scabra, Greene. Scape 6–8 inches high, scabrous, the leaves scabrous — serrulate: perianth segments broad, obtuse or retuse: forks of the filaments slender, erect; anthers white.

T. Hendersoni, Greene. Scape and leaves about 10 inches high: pedicels 6–8, slender, 1–2 inches long: perianth funnel-form, 9 lines long cleft to the middle, yellowish with purple veins: filaments equal, free above the perianth tube: anthers less than a line long, blue, obtuse. Or.

T. hyacinthina, Greene, is in *Bay-Reg. Bot.* **Hesperoscordum lacteum**, Lindl.

BREVOORTIA.

B. coccinea in the key should be **B. Ida-Maia**, Wood. In *Bay-Reg. Bot.* it is **Dichelostemma Ida-Maia**, (Wood) Greene. It was named by Prof. Alfonso Wood for the stage driver's little daughter who was with him when he first saw the flowers, (1866). No matter what generic name is finally fastened upon it by botanists the children of California will call it *Ida May's Fire-crackers.*

Smilacina is **Vagnera** in *Bay-Reg. Bot.* **Maianthemum** is **Unifolium**, and **Prosartes** is **Disporum**, the specific names unchanged in the first and last; **M. bifolium** of this book is **U. dilatatum** in Greene's manual.

FRITILLARIA.

F. coccinea, Greene. Stems 8–18 inches high: leaves in 2 or 3 whorls, linear-lanceolate: flowers 1–4, an inch longer: segments not recurved at tip, yellow and scarlet, checkered: styles distinct above, stigmas linear: pod obtusely angled. Sonoma and Napa Co's.

F. biflora. var. **agrestis**, Greene. Stem 1–2 feet high from an ovoid cluster of sub-cylindraceous bulb-scales: leaves a whorl of 3 below the upper alternate ones: flowers 3–6, nodding-campanulate, greenish, an inch long, odor bad: Stamens shorter than the pistil: style cleft nearly to the base. Interior Coast Range.

. **F. glauca**, Greene. Stems 6–8 inches high: leaves alternate, oblong-lanceolate, glaucous: flowers 1–2, very broadly campanulate, greenish or purplish, the segments an inch long. Near Waldo, Or.

F. mutica in *Bay-Reg. Bot.* is **F. lanceolata**, Pursh., var. **floribunda**, Benth., of this book.

F. pudica, Sprengel. Bulb of 2 or 3 large, rounded scales and many smaller ones: stem 3–8 inches high: leaves 3–8 narrowly oblanceolate or linear, scattered or nearly verticillate, 2–4 inches long, flowers often solitary (1–6), nodding, yellow or orange, often purple-tinged, 5–9 lines long, scarcely spreading.   Yreka, Cal., *Miss Lillian Vivian.*

ERYTHRONIUM.

E. giganteum, Lindl. Leaves 6–10 inches long, narrowed to a short margined petiole; scape 10–15 inches high, 1–6-flowered: perianth cream-color tinged with pink, yellow at base; segments broadly lanceolate, 12–18 lines long, recurved. This is under **E. grandiflorum** in the key (var. **Smithii**) but considered a distinct species by Greene.

E. Hartwegi, Wats. Described by Greene in *Bay-Reg. Bot.* as having pale yellow flowers with orange center and segments scarcely recurved, is, according to him, the species growing about Healdsburg on the Russian River and northward.

E. Hendersoni, Wat. Leaves mottled: scape 1–2 flowered, perianth recurved, 18 lines long, pale purple with darker base fringed with pale yellow; petals auriculate above the short claw, the auricles sub-saccate with globose-inflated scales: slender filaments purple: style clavate, short-lobed stigma cupulate.   Oregon.

E. citrinum, Wat. Similar, flowers smaller, lemon-color with broad orange spots at the base, tips purple-tinged, the filaments yellow or white.   Oregon.

E. Howellii, Wat. Distinguished from the preceding two by pale pink perianth segments with basal orange spots and petals without auricles and scales.

CALOCHORTUS.

C. collinus, Lemmon. Glaucous, 3–10 inches high, simple or branching: flowers 2–6 on pedicils 3–6 inches long: sepals elliptical, greenish yellow: petals creamy white obovate, slightly concave 6–9 lines long; gland cuneate, ciliate, purplish: stamens longer than the pistils: obtuse oblong anthers a line or less long: pods nodding. Near **C. Maweanus.**

C. Lyoni, Wat. Branching stem 1–2 feet high, usually several flowers: sepals naked, acute: petals 12–20 lines long, lilac or purplish, the oblong hairy gland in a larger dark spot: anthers obtuse, 2 lines long or less.   Los Angeles.

C. albus, Dougl, Var. **rubellus,** Greene. Flowers longer and narrower than the type, rose-colored.   Monterey.

ANALYTICAL KEY

TO

GENERA AND SPECIES

OF

WEST COAST PLANTS.

CLASS I.—EXOGENS OR DICOTYLEDONS.

[77]

DIVISION I. POLYPETALÆ.

RANUNCULACEÆ.

* *Petals and sepals similarly colored.*

Sepals and petals slender: carpels 3 to 8 on stipes: smooth; evergreen 8
Petals 5 spur-like sacs: follicles 5: flowers nodding: leaves compound 10
Upper sepal with a spur enclosing spurs of two petals: petals 4 11
Upper sepal a hood enclosing spurs of two petals 12
Minute white flowers in dense capitate racemes: pistil single 13
Flowers pinkish or whitish on scapes 6
Flowers solitary on scapes: receptacle elongated: leaves slender, entire 5

* * *Petals and sepals not similarly colored.*

Petals yellow white or pinkish: akenes small, beaked 6
Petals fleshy, dull purple: follicles large: leaves glaucous 14

* * * *Petals wanting : sepals petaloid.*

Woody climbers: leaves opposite: sepals 4: akenes plumose 1
A whorl of leaves below the flower: akenes many 2
Leaves cordate: sepals greenish, large: follicles 5 to 12 7
Leaves 2-3-ternately compound: follicles pointed 9
Flowers small, greenish, in panicles: leaves 2-3 ternate 3
Flowers small, white; panicles corymbose, akenes 4-angled, inflated 4

1. CLEMATIS, Linnæus.

1. C. ligusticifolia, Nuttall. Leaves 5-foliolate: sepals white, silky.
2. C. lasiantha, Nutt. Leaves 3-foliolate: peduncles 1-2-bracteate.
3. C. pauciflora, Nutt. Leaves fascicled: akenes smooth.
4. C. verticillaris, DC. Flowers bluish purple, large, solitary.

2. ANEMONE, Linnæus.

1. A. occidentalis, Watson. Akenes plumose-tailed: alpine.
2. A. multifida, DC. Akenes densely woolly: sepals villous.
3. A. nemorosa, Linn. Leaves 3, petioled, ternate, incisely lobed.
4. A. deltoidea, Hooker. Leaves usually entire, serrate.

RANUNCULACEÆ.

3. THALICTRUM, Tournefort.

1. T. **polycarpum**, Watson.   Akenes in dense heads, 2 or 3 lines long.
2. T. **occidentalis**, Gray.   Akenes 1 to 6, 3 or 4 lines long, narrower.
3. T. **sparsiflorum**, Turcz.   Anthers obtuse; fruit heads nodding.

4. TRAUTVETTERIA, Fischer & Meyer.

1. T. **grandis**, Nutt.   Slender: leaves few, 5-7-lobed, laciniate-toothed.

5. MYOSURUS, Linnæus.

1. M. **minimus**, Linn.   Receptacle in fruit 1 or 2 inches long.
2. M. **aristatus**, Benth.   Receptacle shorter, akenes beaked.
3. M. **sessilis**, Watson.   Flowers sessile: fruit heads 2 to 6 lines long.

6. RANUNCULUS, Linn.

* *Aquatic: leaves round-reniform and lobed or, when submersed, filiform-dissected.*
Petals white: akenes wrinkled crosswise. 1, 2
Petals yellow: akenes not wrinkled... 6
* * *Not aquatic: leaves all radical: scapes naked or 1-2 bracteate, mostly 1-flowered:*
sepals petaloid.
Sepals white: petals minute: leaves cordate or reniform............................ 3
Sepals pinkish, persistent petals pink: leaves compound........................... 4
* * * *Usually growing in wet places.*
Leaves all entire, oval or narrower..................................... 7, 8, 9, 10
Leaves rather fleshy, simple or 3-foliolate; lobes rounded...................... 5, 12
* * * * *Not aquatic but some species growing in wet places: leaves variously divided*
or lobed.
Alpine, tufted: leaves round-reniform to cuneate, small........................... 11
Some or all the leaves ternately compound: stems branching.
Petals usually more than 6.. 15
Petals usually 5... 13, 16, 17, 18, 19

1. R. **aquatilis**, Linn.   **var. heterophyllus.** Sepals deciduous: receptacle hairy.
2. R. **Lobbii**, Hiern.   Sepals persistent enclosing the few akenes.
3. R. **hystriculus**, Gray.   Akenes 2 or 3 lines long, tapering ; beak hooked.
4. R. **Andersoni**, Gr.   Akenes bladdery, 4 or 5 lines long.
5. R. **Cymbalaria**, Pursh.   Akenes enlarging upward; beak oblique.
6. R. **multifidus**, Pursh.   Petals 5 to 8 with a large scale; beak straight.
7. R. **pusillus**, var. **Lindheimeri**, Gr.   Akenes granulate.
8. R. **Flammula**, Linn. var. **reptans**, Gr. Akenes subglobose.

9. **R. alismæfolius,** Geyer.   Smooth: akenes in globose head.
10. **R Lemmoni,** Gr.   Sepals villous: akenes pubescent.
11. **R oxynotus,** Gr.   Leaves small: sepals hairy: heads oblong.
12. **R. Bloomeri,** Watson.   Akenes straight-beaked.
13. **R. occidentalis.**   Akenes flat, often rough ; beak curved.
14. **R. canus,** Benth.   Densely soft villous when young.
15. **R. Californicus,** Benth.   Petals narrowly obovate, 6 to 20.
16. **R. hispidus,** Michx.   Hispid: calyx scarcely reflexed.
17. **R. orthorhynchus,** Hooker.   Akenes nearly 2 lines long.
18. **R. hebecarpus,** Hook. & Arn.   Flowers minute: akenes bristly.
19. **R. muricatus,** Linn.   Akenes prickly, large, strong-beaked.

7. CALTHA, Linnæus.
1. **C. leptosepala,** D C.   Scape-like stems 1-flowered.

8. COPTIS, Salisbury.
1. **C. asplenifolia.** Salis.   Small petals pouched: sepals slender.
2. **C. occidentalis,** T. & G.   Petals not pouched, linear.   Or.

9. ISOPYRUM, Linnæus.
1. **I. occidentalis,** Hook. & Arn.   Several flowers: follicles 6 lines long.
2. **I. stipitatum,** Gray.   One flower: follicles stipitate, 3 lines long, obtuse.
3. **I. Hallii,** Gray.   Large: 7 to 9 flowers: follicles smaller, acuminate.

10. AQUILEGIA, Tournefort.
1. **A. truncata,** F. & M.   Petals truncate, red, yellow-tinged, Cal
2. **A. formosa,** Fischer.   Similar: petals longer outside.   Oregon.
3. **A. cærulea,** James.   Flowers blue to white: spurs very slender.

11. DELPHINIUM, Tourn.
Flowers blue, purple or white, in terminal racemes.
Capsule pubescent **1, 2, 8, 5**
Capsule glabrous **4, 6, 7, 8**
Flowers red or yellowish in loose racemes. **9, 10**

1. **D. simplex,** Douglas.   Stem and raceme strict: sepals 4 or 5 lines long.
2. **D. variegatum,** T. & G.   More hairy raceme loose: sepals 6 to 10 lines long.
3. **D. Menziesii,** DC.   A longer spur: upper petals purple-veined.
4. **D. decorum,** F & M.   Usually smooth: flowers like the last.
5. **D. depauperatum,** Nuttall.   Smaller: possibly variety of the last.
6

8. **D. Californicum,** T. & G. Dull blue flowers velvety: raceme close.
7. **D. glaucum,** Watson. Glaucous: pale blue flowers: raceme narrow.
8. **D. trollifolium,** Gray. Smooth: leaves shining: flowers large.
9. **D. nudicaule,** T. & G. Follicles narrowed at base: 6 to 12 lines long.
10. **D. cardinale,** Hooker. Follicles broader at base, shorter: flowers larger.

12. ACONITUM, Tournefort.

1. **A. Columbianum,** Nutt. Blue flowers or rarely white: hood beaked.

13. ACTÆA, Linnæus.

1. **A. spicata,** var. **arguta,** Torrey. Leaves 2-3-ternate: berries red.

14. PÆONIA, Linn.

1. **P. Brownii,** Dougl. Leathery sepals persistent: seeds 5 or 6 lines long.

BERBERIDACEÆ.

Low shrubs with spiny-toothed pinnate leaves: flowers yellow........................ 1
Herb: leaves radical, ternately compound: flowers nodding, white................... 2
Herb: radical 3-foliolate leaf solitary: sepals and petals wanting..................... 3

1. BERBERIS, Linnæus.

1. **B. repens,** Lindley. Leaflets 3 to 7, ovate, not shiny: not a foot high.
2. **B. Aquifolium,** Pursh. Leaflets 7 or more, shiny: berries globose.
3. **B. pinnata,** Lag. Petiole short or leaflets at the base: fruit long-ovoid.
4. **B. nervosa,** Pursh. Leaflets palmately nerved, 11 to 17.

2. VANCOUVERIA, Morren & Decaisne.

1. **V. hexandra,** M. & D. Panicle of white flowers on a scape.
Var. **aurea.** Flowers yellow, larger. S. W. Or. (*V. aurea,* Greene.)

3. ACHLYS, De Candolle.

1. **A. triphylla,** DC. Scape ending in a slender spike of minute flowers.

NYMPHÆACEÆ.

Leaves elliptic-peltate, floating: stems jelly-coated............................... 1
Leaves oblong-cordate, large: flowers large, globose, yellow........................ 2

1. BRASENIA, Schreber.

1. **B. peltata, Pursh.** Leaves 2 to 4 inches long: white or purplish flowers.

1. NUPHAR, Smith.

1. **N. polysepalum,** Engelm. Leaves 6 to 12 inches long: stigma broad.

SARRACENIACEÆ.

Darlingtonia Californica, Torrey. Hooded tubular leaves with a pair of mustache-like appendages above the opening: solitary flower nodding on a bracteate scape.

PAPAVERACEÆ.

* *Herbs with entire narrow leaves; the uppermost whorled or opposite: sepals 3, caducous: petals 6 in 2 rows, white or yellowish.*

Filiform stigmas 6 to many; ovaries forming a cylinder............................ 1
Stigmas 3; capsule triangular-ovoid to obovoid or linear........................ 2
* * *Herbs with divided or lobed alternate leaves.*
Sepals 3, winged on the back: half shrubby.................................. 3
Sepals 3 or 2, sharp-horned: bristly with stiff prickles..................... 4
Sepals 2: stigma slightly 4-8-lobed: slender purplish filaments.............. 5
Sepals united into a conical cap: slender stigmas unequal.................... 6
* * * *Shrubs or woody based perennials.*
Sepals 2: buds globular: stigmas 2: leaves entire........................... 7
Sepals 3: petals 6, very large, white: leaves pinnatifid.................... 8

1. PLATYSTEMON, Bentham.

1. **P. Californicus,** Benth. Hirsute: peduncles long, axillary.

2. PLATYSTIGMA, Bentham.

1. **P. lineare,** Benth. Hirsute: stems short: peduncles long: stamens many.
2. **P. Californicum,** B. & H. Capsule 9 to 15 lines long: stamens 10 to 12.
3. **P. Oreganum,** B. & H. Capsule shorter: stamens 4 to 6: smaller.
4. **P. denticulatum,** Greene. Leaves denticulate: stamens 6 to 9.

3. ROMNEYA, Harvey

1. **R. Coulteri,** Harv. White flowers 3 or more inches broad: glaucous leaves

4. ARGEMONE, Linnæus.

1. **A. hispida,** Gray. Densely prickly, petals and stamens only excepted.

5. MECONOPSIS, Viguier.

1. **M. heterophylla**, Benth. Capsule truncate, ribbed, beaked.

6. ESCHSCHOLTZIA, Chamisso.

[Cup-like torus enclosing the ovary 2-margined; the inner membranous, the outer and lower usually thicker (First shown by E. L. Greene)].

    * Outer margin of the obconical torus a broad green or reddish rim.

1. **E. Californica**, Cham. Smooth, perennial, often decumbent.

    * * Torus without conspicuous rim, cylindrical or nearly so.

2. **E. Austinæ**, Greene. Erect, branching, hairy below.
3. **E. tenuifolia**, Benth. Scape-like peduncles square, very slender.
4. **E. rhombipetala**, Greene. Square peduncles rough, stout: petals fugacious.

7. DENDROMECON, Bentham.

1. **D. rigidum**, Benth. Leaves ovate or narrower, rigid, vertical.

FUMARIACEÆ.

Corolla flattened heartshaped or 2-spurred at base............................... 1
Corolla 1-spurred at base, deciduous... 2

1. DICENTRA, Borkhausen.

    * Flowers drooping on a scape : filaments lightly united.

1. **D. formosa**, DC. Raceme com'pd: persistent petals united, rose color.
2. **D. uniflora**, Kellogg. Flower solitary, ½ inch long: capsule short.
3. **D. pauciflora**, Watson. Flowers 1 to 3, 8 to 12 lines long: capsule exserted.
4. **D. cucullaria**, DC. Raceme simple: spurs divergent.

    * * Flowers narrow, erect, in panicles on leafy stems.

5. **D. chrysantha**, H. & A. Petals yellow, 6-9 lines long, tips widely divergent.
6. **D. ochroleuca**, Engelmann. Petals yellowish, longer, tips less divergent.

2. CORYDALIS, Ventenat.

1. **C. aurea**, Willd. var. occidentalis, Gr. Flowers golden yellow.
2. **C. Scouleri**, Hook. Flowers rose-colored in spreading racemes.
3. **C. Caseana**, Gr. Flowers white or cream-color, bluish tipped.
4. **C. Bidwelliæ**, Watson. Similar, but crest entire, spur curved.

CRUCIFERÆ.

§ 1. Pods splitting when ripe, the sides (valves) separating from a central pair of ribs (placentæ) which bear the seeds and usually frame a transparent partition.

* *Pods flattened parallel with the partition, the placental ribs forming the margin: radicle of the bent embryo lying against one edge of the cotyledons (accumbent).*

Pods orbicular, nerveless: flowers small, white or yellowish.

Pods large, flat: seeds orbicular, flat, thin-margined: scapes 1-flowered.............. 1
Pods less than 2 lines broad: flowers in racemes: leaves spatulate, entire............ 2
Pods ovate or lanceolate to linear or oblong not an inch long...................... 3
    Pods narrowly linear, valves nerveless, partition thickened.
Pods long-beaked: stem leaves few, close together near the top..................... 4
Pods short-beaked: leaves scattered: racemes longer.............................. 5
    Pods linear or narrower, an inch or more long, 1-nerved: seeds flat.
Anthers short: petals white, purple or rose-color: claw narrow, blade flat........... 6
Anthers sagittate at base: petals usually unequal and crispate or twisted........... 7
Sepals broad, not colored, the outer gibbous: petals broad, blade flat.............. 8

* * *Pods terete or scarcely flattened, often 4-angled: radicle lying against the side of the cotyledons or embracing them (incumbent), or turned partly to one side (oblique).*

† *Pods slender, 1 to 4 inches long; valves 1-nerved: seeds oblong, slightly flattened; cotyledons often oblique.*

    Flowers white to purple: anthers sagittate.
Petals undulately crisped, little exceeding the large sepals: claw broad............. 9
Petals with flat limb much exceeding the narrow sepals..... 10
    Flowers yellow, large: stem leaves narrow, mostly entire.
Anthers linear, at length coiled: pods on long stipes, curved...................... 11
Anthers sagittate: stigma 2-lobed: pod 4-angled ; no stipe...................... 12

† † *Pods linear, often less than an inch long: seeds in one row (except in 1st sp's. of No. 15 and in No. 16): at least the lower leaves pinnatifid.*

Seeds globose: cotyledons infolding the radicle: anthers sagittate................... 13
Seeds oblong: anthers oblong: leaves lyrately pinnatifid, smooth................... 14
Seeds oblong, small: anthers sagittate: petals 1 to 3 lines long.................... 15
Seeds in two rows: pods 4 to 6 lines long; valves nerveless....................... 16

† † † *Pods oblong-ovoid to globose, beaked with the slender style.*

Densely stellate-pubescent: leaves mostly entire: flowers yellow.................... 17

* * * *Pods flattened contrary to the narrow partition.*

Pods linear, ½ to 2½ inches long, on slender axillary peduncles.................... 18
Pods ovoid, scarcely flattened, on slender scapes.   Aquatic...................... 19
Pods linear, nearly terete: cotyledons 3-parted: petals included................... 20
Pods angular-obcordate or oblong-obovate, many-seeded.......................... 21
Pods oblanceolate to obovate or cuneate-oblong, 4-8-seeded........................ . 22
Pods orbicular to obovate, 2-winged above, 2-seeded.............................. 23

§ 2. Pods not splitting open when ripe: petals minute or wanting except in the last.

Pods with 2 small globular seed-like cells, rough................................ 24

Pods elliptical, twisted, flat, 2 lines long, 6-10-seeded................. 25
Pods minute, orbicular, bristly with hooked hairs, 1-seeded....................... 26
Pods orbicular or obovate, broadly margined, plano-convex, 1-seeded............... 27
Pods terete, spongy-inflated, tapering above, an inch or more long................. 28

1. PLATYSPERMUM, Hooker.

1. **P. scapigerum**, Hook. Glabrous: leaves mostly runcinately lobed. S. N. Mts.

2. ALYSSUM, Tournefort.

1. **A. calycinum**, L. Petals white or yellowish: sepals persistent: pods 4-seeded.
2. **A. maritimum**, L. Petals white: pods 2-seeded (known as Sweet Alyssum).

3. DRABA, Linnæus.

Stems leafy. Nos. 1 to 4. Stems scape-like, few-flowered, not annuals. Nos. 5 to 10
1. **D. cuneifolia**, Nutt. Hirsute, 1 to 6 inches high: petals white, 1 or 2 lines long.
2. **D. stenoloba**, Ledeb. Larger, montane or alpine: petals yellow, obtuse.
3. **D. aureola**, Watson. Densely stellate-hairy: raceme dense: petals yellow.
4. **D. corrugata**, Watson. Pubescence coarser: racemes looser: pod contorted.
5. **D. crassifolia**, Graham. Glabrous: yellow petals a line long: pods acute.
6. **D. Douglasii**, Gr. Glaucous: scapes 6 to 18 lines long: petals white.
7. **D. Lemmoni**, Wats. Stout caudex branching: scapes an inch high: petals yellow.
8. **D. eurycarpa**, Gr. Pod ovate, beaked, 5 to 10 lines long. Sonora Pass.
9. **D. alpina**, L. Petals yellow, 1½ to 2½ lines long. Alpine.
10. **D. Howellii**, Watson. Similar: petals 3 or 4 lines long: pods often one-sided.
N.W. Cal.

4. DENTARIA, Linnæus.

1. **D. tenella**, Pursh. Leaves 1 to 3, 2-5-parted: flowers 3 to 6 lines long.
2. **D. Californica**, Watson. Leaves 2 to 4, toothed, rarely 3-lobed: petals rose-color.

5. CARDAMINE, Linnæus.

Leaves pinnate with several pairs of small leaflets... .... 1, 2, 3
Leaves pinnate with larger leaflets, or simple.
Radical leaves 5-7-foliolate; stem leaves with 5 to 9 entire leaflets............. 4
Radical leaves mostly simple; stem leaves 3-5-foliolate......................... 5, 6
Leaves all ternate, the leaflets 3-5-lobed or toothed: tall...................... 7
Leaves all simple; margin sinuate or entire.................................... 8, 9
1. **C. Gambelii**, Watson. Leaflets 9 to 13, sessile, acute: petals 4 lines long.
2. **C. oligosperma**, Nutt. Leaflets 7 to 11, petiolulate: petals 1 to 1½ lines long.
3. **C. hirsuta**, L. Stouter: leaflets sessile: flowers larger in longer racemes.

4. C. cuneata, Greene. Tuberiferous like the next: leaflets petiolulate.
5. C. paucisecta, Benth. Petals 6 to 9 lines long, white or pinkish.
6. C. Breweri, Watson. Terminal leaflet much the largest: petals 2 lines long.
7. C. angulata, Hook. Flowers few, 3 to 4 lines long: pods short. Oregon.
8. C. cordifolia, Gr. Stout: leaves cordate-orbicular or narrower.
9. C. bellidifolia, L. Alpine, tufted, 2 or 3 in. high: leaves entire.

6. ARABIS, Linnæus.

Pods straight, strictly erect or ascending.
Flowers white, 2 or 3 lines long, in dense elongated racemes...................... **1, 2**
Flowers light pink or rose color, 2 or 3 lines long. Alpine........ .... **3, 4**
Flowers rose-purple, 6 to 9 lines long: leaves dark green, ciliate..................... **5**
Pods curved and usually (except No. 6) more or less reflexed.
Pods 3 in. long, ascending: stout, 2 ft. high, branching............................ **6**
Pods 1 to 4 in. long, strongly reflexed: stem erect, ½ to 2 ft. high................ **7, 8**
Pods 3 or 4 in. long, scarcely a line wide, spreading, recurved..................... **9**
Pods similar, shorter: stem simple, 2 to 10 in. high, villous........................ **10**
1. A. perfoliata, Lam. Glaucous, stout, 2-4 ft. high: leaves crowded clasping.
2. A. hirsuta, Scop. Smaller, more hairy: pods half as long, 1-2 in. Or.
3. A. Lyallii, Watson. Bright green or glaucous, slender: leaves clasping.
4. A. platysperma, Gr. Canescent with stellate hairs: pods 2 lines wide.
5. A. blepharophylla, H. & A. Smooth, often tufted. Coast. Monterey to S. F.
6. A. repanda, Watson. Leaves 3-4 in. long, sinuate toothed: petals 2-3 lines long.
7. A. Holboellii, Hornem. Petals 3-4 lines long, white to purple, reflexed.
8. A. subpinnatifida, Watson. Leaves coarsely toothed: petals pinkish.
9. A. arcuata, Gr. Canescent, hairs branching: petals violet 4-6 lines long.
10. A. Breweri, Watson. Petals deep rose, 1-4 lines long: sepals purplish.

7. STREPTANTHUS, Nuttall.

Glabrous or glaucous: stem-leaves clasping by cordate or sagittate base.
Stem-leaves thick, usually toothed, cordate to narrowly ovate...................... **1**
Stem-leaves rounded cordate, often crowded, entire: pods curved................... **2**
Stem-leaves ovate to lanceolate, acute: pods nearly straight, slender................ **3**
Stem-leaves spatulate: sepals broad, 3 lines long: petal-blades purple.............. **4**
Stem-leaves very narrow, pinnatifid; or some entire, small and cordate.............. **5**
Stem-leaves very slender, margins involute: outer larger sepals subcordate.......... **6**
Glaucous: racemes zigzag: calyx subglobose, black purple........................... **7**
   More or less hispid with simple hairs: flowers purple or red.
Stem-leaves auriculate-clasping, toothed: racemes one-sided..................... **8, 9**
Stem-leaves scarcely clasping: raceme short: flowers often recurved............... **10**

Stem-leaves an inch or less long, not clasping: flowers and pods erect............... 11
1.   S.   cordatus, Nutt.   Petals 4-6 lines long, yellowish to purple.
2.   S.   tortuosus, Kellogg.   Petals similar: pods narrower (a line wide).
3.   S.   Breweri, Gr.   Petals 3-5 lines long, purple: pods 1½ to 2½ in. long.
4.   S.   Howellii, Watson.   Collected in S.W. Or. by Thos. Howell in 1884.
5.   S.   diversifolius, Watson.   Pods strongly reflexed, slender.   Cosumnes River.
6.   S.   polygaloides, Gr.   Sepals yellow: petals purple scarcely exserted.
7.   S.   niger, Greene.   Petals with purple claw and minute veinles white blade.
8.   S.   peramœnus, Greene.   calyx magenta: blade cf petals white, purple-veined
9.   S.   glandulosus, Hook.   Petals red-purple, 6 to 8 lines long.
10.   S.   hispidus, Gr.   Hirsute, 2-5 in. high: red-purple petals 4-6 lines long.
11.   S.   flavescens, Hook.   Petals yellowish, linear: sepals half as long, acute.

8. CHEIRANTHUS, Linnæus.

1.   C.   Menziesii, B. & H.   Smooth stems scape-like: petals purple.   This is *Phœni-caulis Menziesii*, Greene (the generic name given by Nuttall).
2.   C.   asper, C. & S.   Stems erect, leafy: petals orange or yellow.

9. CAULANTHUS, Watson.

1.   C.   procerus, Wats.   Glabrous, 4 to 7 ft. high, stout, branching: flowers greenish.

10. THELYPODIUM, Endlicher.

1.   T.   brachycarpum, Torr.   Stem 1 to 5 ft. high: petals slender, white.   S. N. Mts.
2.   T.   flavescens, Watson   Sepals hairy, yellowish: pod 1½ in. long.
3.   T.   lasiophyllum, Greene.   (*Sisymbrium reflexum* Nutt.)   Leaves pinnatifid: stems 1 to 5 ft. high: pods deflexed or erect.

11. STANLEYA, Nuttall.

1.   S.   pinnatifida, Nutt.   Stems several, 1 to 8 ft. high: petals narrow.

12. ERYSIMUM, Linnæus.

1.   E.   asperum, DC. . Canescent, leafy: petals 8 to 12 lines long.

13. BRASSICA, Linnæus.

1.   B.   nigra, Boiss.   Leaves petioled: pods 4-angled, 6 to 9 lines long.
2.   B.   campestris, L.   Upper leaves clasping: pods 2 or 3 in. long; beak long.
3.   B.   Sinapistrum, Boiss.   Rough-hairy: pods 1 to 1½ in. long, ½ beak.

14. BARBAREA, Robt. Brown.

1.   B.   vulgaris, R. Br.   Perennial, 1 to 3 ft. high: racemes dense, yellow.
Var. arcuata, Koch., has pods and pedicles spreading.   Wet ground.

15. SISYMBRIUM, Linnæus.

Leaves 1-2-pinnate; segments usually pinnatifid: dense racemes: pods acute at both ends,
   3 to 6 lines long, pedicels spreading.. 1, 2
Leaves pinnatifid or entire: pods 10 to 18 lines long............................ 3, 4
Leaves runcinate, 3 to 6 in. long: divaricately branched.......................... 5

1. S. canescens, Nutt. Stems (as in all the species) branching: seeds in 2 rows.
2. S. incisum, Englem. Somewhat glandular: petals 1½ lines long. Montana.
3. S. junceum, Bieb. Glaucous: petals 3 lines long. Oregon.
4. S. · acutangulum, DC. Hairy: leaves runcinate, 2 to 6 in. long.
5. S. officinale, Scop. Similar: pod 6 lines long, tapering to a point.

16. NASTURTIUM, Robt. Brown.

1. N. curvisiliqua, Nutt. Leaves pinnatifid: petals but little exserted.
2. N. obtusum, Nutt. Petals minute: pods 1½ to 3 lines long.
3. N. officinale, R. Br. Aquatic: petals white, 1½ to 2 lines long.

17. VESICARIA, Tournefort.

1. V. montana, Gr. Pods oblong-ovoid, 2½ lines long. N. Cal.
2. V. Kingii, Wats. Leaves 2 to 6 lines long: pods hairy, ovoid.
3. V. occidentalis. Flowers 4 lines long: pods globose. N. Cal.

18. TROPIDOCARPUM, Hooker.

1. T. gracile, Hook. Leaves pinnatifid: flowers yellow, 3 to 6 lines long. Cal.

19. SUBULARIA, Linnæus.

1. S. aquatica, L. Flowers minute: pods 1½ lines long. Mono Pass.

20. STANFORDIA, Watson.

1. S. Californica, Wats. Flowers 3 or 4 lines long on hairy pedicels, purple.

21. CAPSELLA, Moench.

1. C. divaricata, Walp. Very slender, diffuse: pods elliptic oblong.
2. C. Bursa-pastoris, Moench. Pods cuneate-obcordate. Everywhere.

22. THLASPI, Linnæus.

1. T. alpestre, L. Pods obovate to cuneate-oblong, not acute, beaked.
2. T. Californicum, Watson. Pods oblanceolate, acute, 4 or 5 lines long. N. Cal.

CRUCIFERÆ.

23. LEPIDIUM, Linnæus.

1. L. **latipes**, Hook. Stout, rigid stem, 1 to 3 in. long; leaves longer: pods long-winged.
2. L. **dictyotum**, Gr. var. acutidens Gr. Stem slender, 1 to 3 in. high.
3. L. **oxycarpum**, T. & G. Slender: petals none: stamens 2: smooth pods nodding.
4. L. **nitidum**, Nutt. Petals small: pods shining, acutely margined.
5. L. **strictum**. Often matted: sepals persistent: pods erect in dense racemes.
6. L. **Menziesii**, DC. Hispid or pubescent: petals none: pods glabrous.
7. L. **lasiocarpum**, Nutt. Rough-pubescent: pods hispid on margin.
8. L. **Virginicum**, L. Smooth stem erect, at length leafless below, paniculata.
9. L. **Draba**, L. Perennial: leaves not lobed: petals large: pods cordate.
10. L. **campestre**, L. Stout: leaves serrate: pods ovate, broadly winged, scabrous.

24. SENEBIERA, De Candolle.

1. S. **didyma**, Pers. Racemes opposite pinnatifid leaves. Ill-scented.

25. HETERODRABA, E. L. Greene.

1. H. **unilateralis**, Greene. Branching, nearly prostrate; pedicels reflexed.

26. ATHYSANUS, E. L. Greene.

1. A. **pusillus**, Greene. Very slender (*Thysanocarpus pusillus*, Hooker).

27. THYSANOCARPUS, Hooker.

1. T. **curvipes**, Hook. Leaves clasping: border of pods often perforate.
2. T. **laciniatus**, Nutt. More slender: leaves scarcely clasping: pods entire.
3. T. **radians**, Benth. Glabrous: pods with radiating ribs, 4 or 5 lines broad.

28. RAPHANUS, Linnæus.

1. R. **sativus**, L. Petals veiny, color variable: pithy pods 1 to 1½ in. long.
2. R. **Raphanistrum**, L. Similar: pods more constricted between seeds.

CAPPARIDACEÆ.

1. ISOMERIS, Nuttall.

1. I. **arborea**, Nutt. Yellow flowers in bracteate racemes: stamens exserted.

2. CLEOME, Linnæus.

1. C. **platycarpa**, Torr. Erect annual: yellow flowers corymbose; pod hanging.

CISTACEÆ.

HELIANTHEMUM, Tournefort.

H. scoparium, Nutt. Woody-based stems, slender: flowers yellow. Cent. & S. Cal.

VIOLACEÆ.

VIOLA, Linnæus.

* *Leaves all cordate and reniform.*

Stems erect or prostrate: leaves flat............................. .... **5, 10, 11**

Stemless: flowers white or blue... **1, 2**

* * *Leaves not all cordate or reniform: not lobed.*

Flowers blue or violet and white... **3, 4**

Flowers yellow........ .... **7, 8, 9, 10, 12**

* * * *Leaves lobed or divided. Flowers yellow or yellow and purple or blue.*

Stems a few inches to a foot high... **6, 12**

Stems short or none........ .... **13, 14, 15**

1. **V. blanda,** Willd. White flowers. Alpine in wet places.
2. **V. cucullata,** Ait. Light blue to white petals, 5 to 8 lines long.
3. **V. canina,** Linn. var. **adunca** Gr. Blue flowers, long spurred.
4. **V. cuneata,** Watson. Leaves tapering at base: flowers purple and white.
5. **V. ocellata,** T. & G. Leaves coarsely crenate: flowers white, purple marked.
6. **V. Hallii,** Gr. Gray-green: upper petals purple; lower light yellow.
7. **V. pedunculata,** T. & G. Orange-yellow petals brown on the back.
8. **V. præmorsa,** Dougl. A variable species heretofore known as *V. aurea,* Kell. Leaves ovate or narrower, crenate.
9. **V. Nuttallii,** Pursh. Leaves oblong, margins entire. N. Cal., Or.
10 **V. sarmentosa,** Dougl. Slender stems prostrate: leaves small.
11. **V. glabella,** Nutt. Leaves large, bright green, thin, acute.
12. **V. lobata,** Benth. Leaves pedately lobed or some entire.

Var. **integrifolia,** Watson. Leaves not lobed: coarsely toothed.

13. **V. chrysantha,** Hooker. Leaves bipinnatifid: flowers like No. 7.
14. **V. Beckwithii,** T. & G. Flowers purple or blue and yellow.
15. **V. Sheltonii,** Torr. Narrower petals yellow, purple veined.

POLYGALACEÆ.

Side sepals petaloid, larger: petals 3 united with stamens, middle one hooded and beaked.. .... **1**

Sepals and petals 5 each, unequal: stamens 4: fruit prickly.......... .... **2**

1. POLYGALA, Tournefort

1. P. cucullata, Benth.   Flowers rose-color: broad beak obtuse.   Cal.
2. P. Californica, Nutt.   Flowers greenish, purplish: sepals tomentose.   CaL-Or

2. KRAMERIA, Linnæus.

1. K. parvifolia, Benth.   Low rigid shrub: upper petals united.   San Diego.

FRANKENIACEÆ.

1. FRANKENIA, Linnæus.

1. F. grandifolia, C. & S.   Gray-green: calyx-tube furrowed: petals pink.

CARYOPHYLLACEÆ.

* Sepals united : petals long-clawed.
Petals with erect bifid appendages at the base of the blade.

Styles 2 (1 sp. in No. 3).   Styles 3.. 1
Styles 5, rarely 3 or 4: alpine... 2
    Petals not appendaged.   Styles 2... 3
                      Styles 3: petals bifid, white................ Sp. No. 6 of 1
    * * Sepals distinct or nearly so: petals without claws or appendages.
    Stipules none: petals white (or pink in No. 9): rarely wanting.
Petals bifid: pod cylindrical... 4
Petals bifid or wanting pod globular to oblong...................................... 5
Petals entire or wanting: styles opposite sepals................................... 6
Styles not opposite sepals.. 7
    Stipules scarious: pedicel long, reflexed in fruit: leaves not rigid, fascicled, rather
fleshy, filiform to linear: petals entire.. 8
Styles 3, rarely 5 (petals wanting in 3d species)................................... 9
    Stipules scarious: pedicels bracteate or none: stamens 3 to 5: style 3-cleft, or sessile
stigmas 3: petals minute or wanting.
Leaves not rigid: capsule globose... 10
Leaves and sepals rigid: capsule 3-sided.. 11

1. SILENE, Linnæus.

* Blade of the petals entire or only emarginate..

Hairy: pinkish flowers in a 1-sided leafy-spike................................... 4
Glabrous: glutinous rings on the stem: pedicels long.............................. 5
Viscid-pubescent; leafy: pedicels short: petals with 4 appendages................. 21

           .   ** * Blade of the petals bifid or 2-lobed.*

  a. **Segments** or lobes of the petals entire.
  Blade shortly 2-lobed: appendages entire: calyx ovoid or campanulate.
Flowers several, slender pedicelled, brownish purple............................ 2
Flowers solitary, long pedancled... 3
Flowers small, white: no appendages... 6
  Blade cleft to about the middle or deeper, rose color (except 18).
Appendages notched: claw filaments and stipe woolly.......................... 17
Appendages toothed: claw very narrowly auricled, smooth....................... 16
Appendages entire: claw not auricled, smooth: blade rose purple.................. 15
Appendages very small: petal lobes very narrow, white......................... 18
Appendages narrow: claw broadly auricled: petal-lobes broad.................... 19
  b. **Segments** lobed, toothed or notched.
Lobes notched: short appendages toothed: claw not auricled...................... 16
Lobes with a tooth on the outside: claw broad, auricled.......................... 22
Lobes broad: appendages notched: claw auricled................................. 20
Lobes slender, bifid: narrow claws with projecting auricles...................... 14
     * * * *Blade of petals 4-6-parted.*
  Flowers white or pinkish: lobes of the petals mostly filiform.
Calyx open campanulate, nodding: filaments exserted, hairy...................... 1
Calyx ovoid-cylindrical, deflexed in fruit: claws hairy........................... 10
Calyx oblong, erect, much surpassed by the rotate petals........................ 7
Calyx cylindrical, little surpassed by the equally 4-cleft petals.................... 12
Calyx little exceeded by narrow half-inch petals: filaments exserted............... 13
  Flowers scarlet or deep purple, large.
Appendages oblong-lanceolate: claw ciliate: capsule ovoid......................... 8
Appendages ovate: claw smooth: capsule oblong................................. 9
Appendages linear, half as long as the purple blade: claw slightly hairy............. 11

1. S. **campanulata**, Watson. Filiform-disected petals reflexed.
2. S. **Lyalli**, Watson. Stems slender, glabrous: anthers included.
3. S. **monantha**, Watson. Stems weak, elongated. Columbia River.
4. S. **Gallica**, Linn. Rough-hairy: small flowers nearly sessile: annual.
5. S. **antirrhina**, Linn. Glabrous, slender: petals equaling the calyx: annual.
6. S. **Menziesii**, Hooker. Numerous weak stems: flowers small, white.
7. S. **Hookeri**, Nutt. White-tomentose, leafy: erect flowers over an inch broad.
8. S. **Californica**, Durand. Glandular-pubescent: 6 inches to several feet high.
9. S. **laciniata**, Cav. Narrower leaves: petals 4-cleft; segments entire. Cal.
10. S. **Lemmoni**, Watson. Stems many, decumbent, branched: petals white or pinkish.
11. S. **occidentalis**, Watson. Stems erect: petals 4-cleft: stipe 3 lines long.
12. S. **montana**, Watson. Auricles and appendages of petals lacerate.
13. S. **Palmeri**, Watson. Stamens and style much exserted: filaments hairy.

14. S. **Oregana**, Watson. Petals 2-parted, the segments filiform: ovary long-stiped.
15 S. **pectinata**, Watson. Viscid: calyx deeply cleft: petals deep purple.
16. S. **incompta**, Gray. Viscid, tall: lobes of the petals often toothed.
17. S. **verecunda**, Watson. Stems clustered, simple: capsule exserted.
18. S. **Bridgesii**, Rohrb. White petals very narrow: styles long.
19. S. **Douglasii**, Hooker. Similar to No. 17: ovary about equaling calyx.
20. S. **Scouleri**, Hooker. Stout: leaves distant: ovoid capsule, long-stiped. Or.
21. S. **Spaldingii**, Watson. Viscid, leafy: capsule oblong, short-stiped. Or.
22. S. **Grayii**, Watson. Cespitose, grayish: petals and appendages broad. Alpine.

2. LYCHNIS, Tournefort.

1. L. **Californica**, Watson. Petals bifid, lobes on the sides. Alpine. S. N. Mts.

3. SAPONARIA, Linnæus.

1. S. **Vaccaria**, L. Glaucous: calyx 5-angled: entire petals not appendaged. Nat.
2. S. **officinalis**, L. Calyx not angular: petals emarginate, crowned. Nat.

4. CERASTIUM, Linnæus.

1. C. **nutans**, Raf. Viscid, annual: capsule curved, long exserted. San Diego.
2. C. **arvense**, L. Downy, cespitose: capsule nearly straight, short.
3. C. **viscosum**, L. Viscid, annual: leaves broad: capsule straight, long.
4. C. **vulgatum**, L. Leaves narrower: pedicels longer: capsule broader.
5. C. **pilosum**, Ledeb. Flowers ½ in. or more broad: capsule-teeth coiled. Coast.

5. STELLARIA, Linnæus.

Bracts small and scarious or none: leaves acute.
Smooth and shining or glaucous: pedicels erect................................... 1, 2
Glabrous: flowers in umbel-like cyme, long-pediceled............................... 3
Bracts foliaceous: pedicels spreading or deflexed.
Glabrous: petals 2-parted, included or wanting................................... 4
Pubescent, rather stout, 1 or 2 ft. high: petals exceeding calyx.................. 5, 6
Pubescent, spreading: leaves ovate, petioled: petals included....................... 7
1. S. **nitens**, Nutt. Annual: flowers erect; pedicels short: sepals 3-nerved.
2. S. **longipes**, Goldie. Often glancous: leaves stiff: pedicels long.
3. S. **umbellata**, Turcz. Sepals 1-nerved: petals none: capsule exserted.
4. S. **borealis**, Bigelow. Stems weak: pedicels 5-7 lines long: capsule ovoid.
5. S. **Jamesii**, Torr. Viscid: leaves acuminate, long: petals 4-6 lines long.
6. S. **littoralis**, Torr. Leaves ovate, rounded at base: styles rarely 4.
7 S. **media**, Linn. Weak: a hairy line on the stem: petals included, 2-parted.

6. ARENARIA, Linnæus.

* *The 3 valves of the capsule 2-cleft or parted: cespitose perennials with linear-subulate leaves and mostly scarious bracts.*

Petals exceeding the sepals; the capsule about equaling them...................... 1, 2
Petals about equaling the sepals: leaves pungent............................ 3, 4, 5
* * *The 3 valves entire: annuals: bracts leaf-like.*

Much branched: leaves filiform, 3-12 lines long................................. 6, 8
Smooth: leaves lanceolate, obtuse, 1 or 2 lines long.......................... 9, 10
Leaves linear to lanceolate 6-12 lines long.................................. 7, 11
* * * *Parts of the flower sometimes in 4's: capsular valves bifid: leaves bright green, 1 or 2 inches long...* 12, 13

1. A. congesta, Nutt. Glaucous: flowers in dense fascicles: bracts broad.
2. A. capillaris, Poir. Pubescent: flowers few: bracts small, lanceolate.
3. A. pungens, Nutt. Stems 2 or 3 inches high, leafy, pubescent. Subalpine.
4. A. Franklinii, Dougl. Stouter: sepals shining, margin scarious. Or.
5. A. verna, Linn. Leaves erect, 2-3 lines long: sepals exceeding petals.
6. A. Douglasii, T. & G. Capsule globose: seeds flat, smooth.
7. A. Howellii, Watson. Glandular-hispid, a foot high. Or.
8. A. tenella, Nutt. Capsule oblong: seeds rough: sepals 3-nerved. Or. N.
9. A. Californica, Brewer. Sepals 3-nerved: Seeds rough. Cent. Cal.
10. A. pusilla, Watson. Sepals 1-nerved: petals minute or none: seeds smooth.
11. A. palustris, Watson. Stems simple: leaves flaccid: few pedicels long.
12. A. macrophylla, Hooker. Leaves acute, 3 or 4 pairs: petals obtuse.
13. A. lateriflora, Linn. Leaves broader, obtuse: petals exserted. Or.

7. SAGINA, Linnæus.

1. S. occidentalis, Watson. Slender: 2-6 inches high: pedicels straight.
2. S. Linnæi, Presl. Densely matted: 1-2 in. high: fruiting pedicels curved.
3. S. crassicaulis, Watson. Stout, branched: leaves fleshy, scarious at base.

8. SPERGULA, Linnæus.

1. S· arvensis, L. Leaves filiform, smooth: sepals and petals equal.

9. LEPIGONUM, Fries.

1. L. macrothecum, F. & M. Pubescent: sepals and petals 3 lines long or more.
2. L. medium, Fries. More slender: flowers smaller, white.
3. L. gracile, Watson. Annual, smooth, slender: sepals ½-1 line long.

10. POLYCARPON, Linnæus.

1. P. depressum. Nutt. An inch high: petals included, entire.

11. LŒFLINGIA, Linnæus.

1. L. squarrosa, Nutt. Glandular-pubescent: 2-6 inches high.

ILLECEBRACEÆ.

PENTACÆNA, Bartling.

1. P. ramosissima, H. & A. Prostrate: subulate pungent gray-green leaves crowded: stipules silvery: sessile flowers clustered: sepals 5, hooded, ending in a spine.

PORTULACACEÆ.

* *Sepals 2, united below and adherent to the partly inferior ovary.*

Flowers yellow or rose-red: capsule opening by a lid............................... 1

  * * *Sepals 2, persistent, not adherent to the superior ovary.*

Style 3-cleft (rarely 2-cleft in Calandrinia): sepals green.

Stamens more than 6: petals 5 or more... 2

Stamens 5 (3 in No. 8): petals 5.. 3

Stamens 3: petals 5, unequal, coherent: leaves or leaf and bract opposite............ 4

Style 2-cleft: sepals membranous rounded-cordate exceeding the 2 or 4 petals........ 5

* * * *Sepals 4 to 8, distinct, unequal, persistent.*........................... 6

1. PORTULACA, Tournefort.

1. P. oleracea, L. Leaves obovate to spatulate: petals yellow, 1 to 2 lines. Nat.

2. P. pilosa, L. Leaves linear: nearly terete: petals red, 2 or 3 lines long.

2. CALANDRINIA, HBK.

Leafy stems, annual: flowers in racemes: petals 3 to 5, rose-red............... 1, 2

Leaves mostly all radical: perennial: sepals orbicular.

Leaves linear, all radical: short scape 2-bracteate. Alpine...................... 3, 4

Leaves oblanceolate to obovate, all radical (except No. 5)..................... 5, 6, 7

1. C. Menziesii. Hooker. Sepals keeled, acute: capsule ovoid, acute.

2. C. Breweri, Watson. Capsule longer, conical, obtuse on deflexed pedicels.

3. C. pygmæa, Gray. Bracts scarious: sepals dentate: petals red.

4. C. Nevadensis, Gr. Larger: bracts green: sepals entire: petals white.

5. C. oppositifolia, Watson. Stem with 2 or 3 pairs of opposite leaves: petals white.

6. C. cotyledon, Wat. Scape with lanceolate ciliate bracts: petals rose-red.

7. C. Leana, Porter. Similar: petals 6 to 8. N. Cal. Or.

3. CLAYTONIA, Linnæus.

a. Annuals with fibrous roots, rarely with bulblets.

Stems simple, bearing a single pair of united or distinct leaves.

Leaves united into a cup enclosing the raceme... 1

Leaves united at the base on one or both sides................................. 2, 8

Leaves distinct (No. 4 with bulblets at base)..................... 2, 3, 4, 5, 11, 13

Stems usually branching, leafy.

Leaves opposite stems often rooting at joints and bulbiferous........................ 6

Leaves alternate... 7, 8, 9, 10

b. Perennials, with deep-seated tubers, stem leaves, a pair or a whorl.. 11, 12, 13

1. C. perfoliata, Donn. From 1 to 12 inches high: radical leaves broad.

Var. parviflora, Torr. Radical leaves all linear or spatulate.

2. C. spathulata, Dougl. Very slender: leaves distinct or united on one side.

3. C. exigua, T. &. G. Glaucous: leaves nearly filiform; the pair broader, united at base.

4. C. bulbifera, Gr. Stems lax; long pedicels with conspicuous bracts.

5. C. cordifolia, Watson. Pair of leaves ovate, acute; radical, cordate: no bracts.

6. C. Chamissonis, Esch. Leaves oblanceolate: petals white.

7. C. parviflora, Mocino. Very slender: leaves broadly spatulate, small.

8. C. dichotoma, Nutt. Small: leaves linear: petals unequal: stamens 3.

9. C. linearis, Dougl. Leaves slender, clasping: sepals broad, often colored: petals white.

10. C. diffusa, Nutt. Leaves ovate or deltoid, petioled: racemes often axillary.

11. C. Caroliniana, Michx., var. sessilifolia, Torr. Usually one radical leaf; the pair lanceolate to linear: a single scarious bract.

12. C. triphylla, Watson. Leaves slender: raceme compound, bracts scarious.

13. C. Nevadensis, Watson. Leaves ovate to orbicular: petals 3-5 lines long, clawed.

4. MONTIA, Linnæus.

1. M. fontana, L. Stems weak, often matted: flowers minute: capsule globose.

2. M. Howellii, Watson. Leaves opposite the scarious triangular bracts of racemes.

5. CALYPTRIDIUM, Nuttall.

1. C. umbellatum, Greene. Umbel capitate: petals 4: stamens 3: style exserted. This plant is *Spraguea umbellata*, Torr.

2. C. quadripetalum, Watson. Petals 4: stamen 1: stigmas nearly sessile.

3. C. roseum, Watson. Petals 2, much shorter than the unequal sepals.

4. C. monandrum. Nutt. Petals 2, equalling the sepals, a line long or less.

6. LEWISIA, Pursh.

1. **L. rediviva**, Pursh. Scapes with a whorl of scarious bracts.
2. **L. brachycalyx**, Engelm. Scapes 2-bracted at base: sepals 4.

ELATINACEÆ.

Small prostrate aquatics with entire leaves: parts of the flower each 2 to 4 (except in
    sp. No. 3).   Sepals obtuse: membranaceous.................................... 1
Erect, glandular-pubescent: parts of the flower in 5's............................. 2

ELATINE, Linnæus.

1. **E. Americana**, Arnott.   Seeds pitted in 9 to 10 lines, ½ line long.
2. **E. brachysperma**, Gray.   Seeds pitted in 6 or 7 lines, shorter.
3. **E. California.**   Flowers not sessile: seeds much curved: stamens 6 to 8.

BERGIA, Linnæus.

1. **B. Texana**, Seubert.   Leaves serrulate: flowers fascicled.

HYPERICACEÆ.

HYPERICUM, Linnæus.

Stamens very numerous: styles 3, long... 1, 2
Stamens 15 to 20: styles 3, short: petals included............................... 3
1. H. formosum, HBK. var. Scouleri, Coulter.   Flowers 6 lines broad. Wet ground.
2. H. concinnum, Benth.   Leaves acute: flowers an inch broad.   Dry ground.
3. H. anagalloides, C. & S.   Leaves 2-6 lines long: flowers 3-4 lines broad.   Wet
    ground.

MALVACEÆ.

    Column of stamens bearing anthers at the topt carpels in a ring around the axis.
Calyx-bracts 2 or 3, united below: an evergreen ever-blooming shrub................ 1
Calyx-bracts 3, distinct: flowers axillary, pinkish: leaves 5-7-lobed................. 2
Calyx-bracts none: flowers racemose or spicate. 3
Calyx-bracts 1 to 3 or none: densely tomentose (except in sp. 6 & 7)........... 4
Calyx-bracts 1 or 2, slender: leaves 1-sided: flowers yellowish...................... 5
    Column of stamens naked at top, 5-toothed: carpels forming a many-seeded capsule
Calyx-bracts many................................ 6

1. LAVATERA, Linnæus.

L. assurgentiflora, Kellogg. Showy flowers in axillary clusters.

2. MALVA, Linn.

M. rotundifolia and M. borealis are introduced weeds.

3. SIDALCEA, Gray.

* *Perennials with usually clustered stems decumbent at base.*

Raceme loose: no stellate hairs: rose-purple petals an inch long....................... 1
Raceme spicate: simple and stellate hairs: petals notched, pinkish, 6 lines long........ 2
Like No. 2, but the larger flowers deep lilac-purple................................. 3
Stems branching: calyx globose in fruit: carpels smooth, straight.................... 4
Nearly glabrous, glaucous, pale, decumbent: petals obtuse or truncate............... 5
Stellate pubescence short: large leaves dark green, slightly 5-lobed.................. 6

* * *Annuals with erect branching stems.*

Carpels strongly incurved and sharply rugose on back............................. 7
Carpels not incurved or rugose, conspicuously hairy-beaked........................ 8
Carpels several—nerved along the back: calyx-lobes abruptly acuminate............. 9
Pedicels subtended by 5-7-parted hispid bracts: calyx-lobes slender................10
Large, with cordate, 3-7-angled leaves: white flowers in close clusters:..............11

1. S. malvæflora, Gr. (S. humilis, Gr. of Cal. Bot., etc.) Common coast species.
2. S. spicata, Greene. Carpels small, hairy, not reticulated. Sierra Nevada.
3. S. campestris, Greene. Stems bristly with deflexed hairs: calyx stellate-hairy.
4. S. Oregana, Gr. Glabrous below, 1 to 5 ft. high: corolla 6 lines long or more.
5. S. glaucesens, Greene. Calyx lobes very slender. High Sierras.
6. S. asprella, Greene. Decumbent, leaves shaped alike. Foot Hills, Sierras.
7. S. Hartwegi, Gr. Glabrous except the hispid calyx and pedicels. Sac. Val.
8. S. hirsuta, Gr. Stout and tall, branching: flower-clusters dense. Chico, Cal.
9. S. calycosa, Jones. Corolla small, light purple: calyx long-ciliate. Cent. Cal.
10. S. diploscypha, Gr. Hirsute: flowers large, umbellate clustered. Cent. Cal.
Var. minor, Gr. Flowers racemose: petals with a spot at base. Cent. Cal.
11. S. malachroides, Gr. Petals obcordate: carpels smooth. Redwoods, Cal. Coast.

4. MALVASTRUM, Gray.

Perennial, often shrubby: stems hoary or gray with soft pubescence............ 1 to 5
Perennial: densely stellate-hairy or hispid: dense-flowered 6
Annual erect with spreading hairs: leaves reniform, long petioled................... 7
Annual, decumbent: small leaves 5-lobed: flowers mostly solitary................... 8
1. M. Thurberi, Gr. Shrubby, branches slender: spikes naked: flowers small.

2. **M. Fremonti,** Torr. Similar: calyx globose in fruit, very woolly.
3. **M. splendidum,** Kellogg. Tall shrub: flowers in large panicles, rose-red. S. Cal
4. **M. marrubioides,** D. & H. Low: leaves serrate, thick: calyx-lobes slender. C.Cai
5. **M. Palmeri,** Watson. Densely stellate-pubescent: large flowers yellowish.
6. **M. densiflorum,** Watson. Hispid bracts very long: calyx-lobes long attenuate.
7. **M. rotundifolium,** Gr. Low: petals 6 lines long, a red spot at the base. S. Cal
8. **M. exile,** Gr. Pedicels slender: petals obovate, 2 to 5 lines long. S. Cal.

5. SIDA, Linnæus.

1. **S. hederacea,** Torr. Decumbent: leaves 1-sided: petals yellowish.

6. HIBISCUS, Linnæus.

1. **Californicus,** Kellogg. Flowers axillary, white with purple, large.

STERCULIACEÆ.

FREMONTIA, Torrey.

1. **F. Californica,** Torr. Tall shrub: flowers yellow, axillary, apetalous. S. N. Mts

LINACEÆ.

LINUM, Linnæus.

Styles 2: flowers yellow: leaves opposite, glabrous, oblong........•................. 1
Styles 3: flower yellow: leaves alternate.. 2, 3
Styles 3: flowers white to rose-color: leaves alternate (or whorled in 7).
     Petals with 3-parted or 3-lobed appendage at base......'.................. 5, 6, 7
     Petals 2-toothed at base, scarcely longer than the sepals..................... 8
Styles 5: flowers large, blue: leaves alternate................................. 9, 10

1. **L. digynum,** Gr. Sepals denticulate, a line long. Near Yosemite Valley. N. Cal.
2. **L. Breweri,** Gr. Glaucous: leaves small, very slender, basal glands large.
3. **L. adenophyllum,** Gr. Leaves margined with stipitate glands. Cent. Cal.
4. **L. Californicum,** Benth. Glaucous: petals 4 lines long: capsule acute.
5. **L. congestum,** Gr. Calyx pubescent: flowers in close clusters. S. F. Bay.
6. **L. spergulinum,** Gr. No stipular glands: petals 2-3 lines long. S. F. Bay.
7. **L. drymarioides,** Curran. Pubescent: leaves ovate, margins glandular. C. Cal
8. **L. micranthum,** Gr. Flowers minute: capsule exserted. Mts. Cal.
9. **L. perenne,** Linn. Perennial, glaucous: flowers large, blue.
10. **L. usatissimum,** L. Similar but annual. The common cultivated flax.

GERANIACEÆ.

Carpels 5, 1-seeded, separating with styles when ripe from the long axis.
 Fertile stamens 10: tails of carpels coiled, not bearded............................ 1
 Fertile stamens 5: tails of carpels twisted, bearded............................. 2
Carpels 5, 1-seeded, fleshy, globular: stamens 10.................................. 3
Carpels united into a 5-celled ovary: capsule 5-sided.............................. 4

1. GERANIUM, Linnæus.

1. G. Carolinianum, L. Petals 2 or 3 lines long. A common weed.
2. G. incisum, Nutt. Flowers deep rose-purple, an inch broad.

2. ERODIUM, L'Heritier.

1. E. cicutarium, L'H. Pinnate leaves: leaflets pinnatifid. (" Filaree.")
2. E. moschatum, L'H. Leaflets doubly toothed: musky.
3. E. Botrys, Bertoloni. Leaves oblong, pinnatifidly lobed. Cent. Cal.
4. E. macrophyllum, H. & A. Leaves palmately lobed. Cent. & S. Cal.

3. LIMNANTHES, Robt. Brown.

1. L. Douglasii, R. Br. Glabrous: petals yellow, white tipped. Cal.
2. L. rosea, Hartweg. Glabrous: petals purple tinged, obovate. Sac. Val.
3. L. alba, Hartweg. Pubescent: petals white or nearly so. Cal.

4. OXALIS, Linnæus.

1. O. Oregana, Nutt. Flowers pinkish. In coast forests.
2. O. trilliifolia, Hook. In Oregon (?).
3. O. corniculata, L. Slender branching stems: flowers yellow.

RUTACEÆ.

A tall shrub or tree: leaves 3-foliolate: flowers in terminal clusters................... 1
A low shrub: leaves simple, opposite: flowers axillary............................... 2

1. PTELEA, Linnæus.

1. P. angustifolia, Benth. Fruit broadly winged, orbicular. Cent. Cal.

2. CNEORIDIUM, Hook. f.

1. O. dumosum, H. f. Leaves often fascicled: fruit drupe-like. San Diego.

CELASTRACEÆ.

A slender deciduous shrub with 4-angled branches: leaves 2-4 inches long............ 1
A low much-branched evergreen: leaves 6-18 lines long, numerous................... 2

1. EUONYMUS, Tournefort.

ι. E. occidentalis, Tourn. Flowers dark brown, parts in 5's, rotate, drooping.

2. PACHYSTIMA, Rafinesque.

1. P. Mersinites, Raf. Flowers greenish, parts in 4's, about a line broad.

RHAMNACEÆ.

Flowers greenish. Leaves alternate: flexuose branches spiny........................ 1
              Leaves alternate: not spiny: fruit juicy......................... 2
              Leaves opposite, 1 or 2 lines long: fruit dry.......... 3
Flowers white or blue, in dense clusters: fruit dry............................... 4

1. ZIZYPHUS, Jussieu.

ι. Z. Parryi, Torr. Peduncles axillary, recurved in fruit, 1-3-flowered.

2. RHAMNUS, Linnæus.

Flowers apetalous and mostly diœcious: seeds concave........................... 1, 2
Flowers with minute petals, mostly perfect: seeds convex on the back............ 3, 4
1. R. alnifolia, L'Her. Low: deciduous leaves acute at each end, serrate.
2. R. crocea, Nutt. Leaves acutely denticulate, evergreen, thin.
3. R. Californica, Esch. Leaves elliptical to ovate-oblong, evergreen, thick.
Var. tomentella, Gr. Densely white-tomentose. Both forms common in Cal.
4. R. Purshiana, DC. Elliptical leaves 2 to 7 inches long, deciduous.

3. ADOLPHIA, Meisner.

ι. A. Californica, Watson. In dense clumps 2 or 3 ft. high, branchlets spiny.

4. CEANOTHUS, Linnæus.

§ 1. Leaves all alternate: fruit not crested.
Leaves 3-nerved from the base.
    Branches not rigid or spiny: leaves glandular serrate (except No. 1).
       Flowers white in large clusters..... .... 1, 2, 3
       Flowers blue... 4, 5, 6

Branches rigid, spreading, often spinose: racemes simple.

Leaves glandular-serrate: flowers blue............................... **7, 8**

Leaves usually entire: branches grayish........................ **9, 10, 11**

Leaves pinnately veined, obtuse: flowers blue (see No. 5)............ **12, 13, 14, 15**

§ 2. Leaves small, often opposite, very thick with numerous straight side veins, spinosely toothed or entire: stipules mostly large and warty: flowers in sessile or shortly peduncled axillary clusters: fruit with 3 projections: branches rigid;.......... **16 to 20**

1. C. **integerrimus**, H. & A. Slender branches round: leaves thin.

2. C. **velutinus**, Dougl. Stout: leaves thick, resinous above.

3. C. **sanguineus**, Pursh. Branches reddish: leaves thin; petioles slender.

4. C. **thyrsiflorus**, Esch. Branches angled: leaves shining above, ashy beneath.

5. C. **dentatus**, T. & G. Leaves mostly 3-4 lines long, thin: thyrse globose.

6. C. **decumbens**, Watson. Trailing, hirsute: leaves thin, teeth green-glandular.

7. C. **hirsutus**, Nutt. Silky, rarely spiny: leaves rounded or cordate at base, acute.

8. C. **sorediatus**, H. & A. Leaves smooth above: racemes pubescent, peduncles short.

9. C. **divaricatus**, Nutt. Branches sometimes green: racemes 1-4 inches long.

10. C. **incanus**, T. & G. Leaves hoary beneath: racemes short: fruit warty.

11. C. **cordulatus**, Kell. Pubescent, low, flat-topped: racemes an inch long or less.

12. C. **spinosus**, Nutt. Often a tree: leaves entire, oblong, thick; petioles slender.

13. C. **papillosus**, T. & G. Leaves narrow, dark green, shining and pimply above.

14. C. **floribundus**, Hooker. Leaves 3-4 lines long, acute, undulate, denticulate.

15. C. **Veitchianus**, Hooker. Glabrous, leaves thick, obovate-cuneate. Rare.

16. C. **crassifolius**, Torr. Branches hoary: leaves tomentose beneath. Cal. Coast.

17. C. **cuneatus**, Nutt. Bark ashy gray: leaves cuneate-obovate, entire. Common.

18. C. **macrocarpus**, Nutt. Tree-like, 8 to 12 ft. high: fruit very large. St. Barbara.

19. C. **rigidus**, Nutt. Branchlets tomentose: leaves 2 to 5 lines long: flowers blue.

20. C. **prostratus**, Benth. Prostrate: leaves spinose at apex only: flowers blue.

VITACEÆ.

VITIS, Tournefort.

1. V. **Californica**, Benth. Leaves round-cordate, serrate. (Wild Grape.)

SAPINDACEÆ.

Flowers in large terminal erect thyrses: calyx tubular: clawed petals unequal........ **1**

Flowers small, the fertile ones in drooping clusters: ovary 2-lobed: fruit 2-winged.

Leaves palmately lobed... **2**

Leaves pinnately 3-foliolate.. 3
Flowers in drooping racemes: stamens 5, much exserted: leaves 3-foliolate, serrulate... 4

1. ÆSCULUS, Linnæus.

1. **Æ. Californicus,** Nutt. Leaves palmately 4-7-foliolate. (Buckeye.)

2. ACER, Tournefort.

1. **A. macrophyllum,** Pursh. Yellowish flowers in dense racemes: fruit hairy.
2. **A. circinatum,** Pursh. Corymbs 10-20 flowered: sepals red or purplish.
3. **A. glabrum,** Torr. Sepals and petals greenish yellow: filaments naked.

3. NEGUNDO, Mœnch.

1. **N. Californicum,** T. & G. Calyx minute: petals none, diœcious. (Box-Elder.)

4. STAPHYLEA, Linnæus.

2. **S. Bolanderi,** Gr. Leaflets broad, stipellate: fruit bladdery. Shasta.

ANACARDIACEÆ.

RHUS, Linnæus.

Slender deciduous shrubs: leaves 3-foliolate: fruit compressed globose.
     Flowers in dense axillary panicles: fruit smooth, dry, whitish................. 1
     Flowers in short scaly-bracted spikes: fruit hairy, gummy, scarlet............. 2
Stout, diffuse evergreen shrubs: leaves simple, coriaceous: fruit ovoid.
     Flowers rose-color: leaves ovate on short petioles............................. 3
     Flowers yellowish: leaves lanceolate on slender petioles....................... 4

1. **R. diversiloba,** T. & Gr. Stems erect or climbing by rootlets (Poison Oak).
2. **R. aromatica,** Ait. **var. trilobata,** Gr. Diffusely slender-branched.
3. **R. integrifolia,** B. & H. Leaves entire or spinosely-toothed: fruit red, frosty.
4. **R. laurina,** Nutt. Leaves glaucous, entire: panicles 2-4 inches long.

LEGUMINOSÆ.

§ 1. Stamens distinct: shrubs (except No. 1).
Leaves palmately 3-foliolate: yellow flowers in terminal close racemes............... 1
Stiff, much branched, evergreen: flowers red-purple, solitary, axillary............,..... 2
Leaves pinnate: flowers purple in dense axillary spikes: petal 1..................... 9

Leaves simple, entire, cordate: flowers rose-purple in axillary clusters.......... 14
§ 2. Stamens all united or one above distinct: herbs (except some in 3 & 7).
* *Leaves palmate with more than 3 leaflets: flowers in heads or racemes.*
Leaflets entire. Spikes or racemes terminal: anthers of 2 kinds............... 3
        Yellow flowers 1 to 5 with a bract...................Sp. No. 10 in 7
        Purplish flowers: stipules not adnate, deciduous................... 8
Leaflets toothed or entire: stipules adnate: anthers alike.....................§ 1 in 4
    * * *Leaves 3-foliolate, palmate or pinnate.*
Leaves palmate: flowers in heads or short spikes: corolla persistent.................. 4
Leaves pinnate: flowers in axillary spikes or racemes.
    Corolla yellow or white: pod wrinkled: leaves fragrant........................ 5
    Corolla purple or greenish: leaflets entire: stipules free....................... 8
    Corolla yellow or purple: leaflets toothed: pod curved or coiled................ 6
Leaves pinnate. Flowers in small axillary clusters, yellow: pod spiral, prickly....... 6
    Flowers solitary or in wheel-like clusters, axillary................. 7
    * * * *Leaves pinnately 4-many-foliolate with a terminal leaflet.*
Flowers solitary or in umbellate whorls, axillary.................................... 7
Flowers in axillary spikes: pod prickly: leaves sticky..............................10
Flowers in axillary spikes or heads: pod often inflated, often 2-celled...............11
    * * * * *Leaves pinnate, ending in a bristle, imperfect leaflet or a tendril.*
Style filiform, hairy around the apex..12
Style flattened, usually twisted half around, one side hairy........................13

1. THERMOPSIS, Robt. Brown.

1. T. **Californica**, Watson. Short-woolly: pod 6-8-ovuled, stipe short.
2. T. **montana**, Nutt. Rather silky: leaflets smooth above: pod 10-12- seeded
3. T. **macrophylla**, H. & A. Villous: leaves oblong-elliptical acute: seeds 4 or 5.

2. PICKERINGIA, Nuttall.

1. P. **montana**, Nutt. Leaves 1-3-foliolate, numerous: stamens persistent.

3. LUPINUS, Linnæus.

A. *Perennials, more or less shrubby, leafy, silky: ovules 6 to 12*............. 1, 2, 3, 4
B. *Perennial herbs, mostly tall; flowers large; bracts deciduous: ovules 6 or more.*
Woody at base: silky: calyx-lips nearly equal................................. 1, 2, 3
Stems mostly stout and hollow: leaflets glabrous above........................ 5, 6
Stems slender, not erect: leaflets an inch long or less............................ 7
Stems leafy and branching: petioles and bracts short...................... 8, 9, 10
C. *Perennial herbs: flowers small: (Ex. No. 13): not yellow: ovules 3 to 6.*
Leaves distant, not glabrous above; lower petioles long: keel ciliate.......... 11 to 16

Leafy: petioles and peduncles mostly short: bracts deciduous: ovules 3 to 5.. 17, 18, 19

D.   *Dwarf alpine perennials, mostly tufted: lower calyx-lip 3-toothed: keel ciliate: pod hairy, 3-4- seeded* .... 20 to 23

E.   *Annuals: leaflets mostly 5 to 7 (8 to 10 in No. 29): bracts falling with or before the petals: upper calyx-lip 2-parted or bifid: pod 4-8-seeded.*

Bracts deciduous: flowers in whorls, 5 or 6 lines long 24, 25

                                2 or 3 lines long .... 26 to 28

Bracts deciduous or persistent for a while: flowers scattered .... 29 to 35

F.   *Annuals: leaflets cuneate-oblong or obovate: bracts conspicuous, persistent in fruit: ovules and seeds 2* .... 36 to 38

1.   **L.** arboreus, Sims. A shrub: flowers yellow, rarely purplish, fragrant. Cal.

2.   **L.** Chamissonis, Esch. A low shrub: flowers blue, rarely violet, pink or white.

3.   **L.** Douglasii, Agardh. Woody at base: much like forms of the last. Cal.

4.   **L.** Ludovicianus, Greene. Shrubby: very villous: flowers purple: pod 5-seeded

5.   **L.** polyphyllus, Lindl. Leaflets numerous, large: raceme purple, long.

6.   **L.** rivularis, Dougl. Stipules very slender: leaflets 7 to 10: petioles short.

7.   **L.** littoralis, Dougl. Leaflets 6 to 12 lines long: racemes short: ovules 10 to 12.

8.   **L.** Sabinii, Dougl. Stipules long, setaceous: flowers bright yellow. Blue Mts.

9.   **L.** albicaulis, Dougl. Reflexed margin of the acute standard coherent at apex.

10.   **L.** ornatus, Dougl. Standard silky: keel ciliate: stipules setaceous: seeds white.

11.   **L** sericeus, Pursh. Bracts long: calyx densely silky, gibbous: pod densely hairy.

12.   **L.** leucophyllus, Dougl Densely silky: dense racemes sessile: standard hairy

13.   **L.** Grayi, Watson. A span high: very hoary-tomentose: racemes short, loose.

14.   **L.** lepidus, Dougl. Low, slender, silky: peduncle and raceme long: petals violet.

15.   **L.** confertus, Kell. Similar but bracts persistent: corolla blue to rose.

16.   **L.** onustus, Watson. Decumbent woody base: flowers deep blue, scattered.

17.   **L.** parviflorus, Nutt. Stems slender, 2 or 3 ft. high, strict, glabrous above.

18   **L.** Andersoni, Watson. Appressed-pubescent, much branched: racemes short.

19.   **L.** laxiflorus, Dougl. Silky: raceme slender: calyx saccate or spurred.

20.   **L.** aridus, Dougl. Raceme dense, 2 or 3 inches long: peduncle short: petals purple.

21.   **L.** minimus, Dougl. Similar, more silky: peduncles longer: standard broader.

22.   **L.** Breweri, Gr. Stems from spreading woody base: densely silky: leaflets obovate.

23.   **L.** Lyallii, Gr. Similar: petioles longer: standard narrower: petals violet. Or.

24.   **L.** affinis, Agardh. A foot high: leaflets broadly obovate: bracts short. Cal.

25.   **L.** nanus, Dougl. Slender: bracts long: petals broad, purple and white. Cal.

26.   **L.** micranthus, Dougl. Slender, branched, decumbent, villous: racemes short.

27.   **L** trifidus, Torr. Similar: lower calyx-lip 3-cleft: pod 5-6-seeded. San F'co Bay.

28.   **L** citrinus, Kell. Similar: calyx-lip 3-toothed: flowers orange or yellow. Fresno.

29.   **L.** leptophyllus, Benth. Bracts very long: standard with a crimson spot. Cal.

30.   **L.** sparsiflorus, Benth. Similar: bracts short, persisting longer: petals violet.

31.   **L.** truncatus, Nutt. Linear leaflets truncate or 3-toothed, smooth above: petals

purple. This and the last two in Cent. Cal., southward.

32. L. **Stiveri**, Kell. Leaflets broad: petioles short: standard yellow: wings rose.
33. L. **hirsutissimus**, Benth. Very hispid with viscid stinging hairs: petals purple.
34. L. **concinnus**, Agardh. Very villous: lower calyx-lip trifed: standard with yellow.
35. L. **gracilis**, Agardh. Leaflets broad, 3 to 6 lines long: petals 2 or 3 lines long. blue and white. Monterey, S. Rare.
36. L. **microcarpus**, Sims. Calyx very villous: flowers usually blue or purple.
37. L. **densiflorus**, Benth. Calyx only finely pubescent: flowers usually yellowish.
38. L. **luteolus**, Kell. Leaves scattered: petioles short: flowers pale yellow. Cal.

4. TRIFOLIUM, Linnæus.

Leaflets mostly 5 to 7... ... 1 to 4
Leaflets 3: heads with no involucre.
   Flowers white or yellowish: leaflets linear to oblong................. 5, 6, 7, 12
   Flowers red, 6 lines long or more........................... .... 8, 9
   Flowers small, at length reflexed..................... 10, 11, 12, 14, 15, 16
Leaflets 3: heads with an involucre: peduncles axillary.
   Involucre green, rotate, the lobes laciniately toothed................... 17 to 20
   Involucre with entire lobes.. 21, 25, 26
   Involucre cup-shaped or broad, lobes toothed....................... 22, 23, 24
   Involucre very small or reduced to a ring............................... 26

§ 1. *Leaflets 5 to 7, rarely 3: calyx teeth filiform, plumose. Alpine perennials.*

1. T. **megacephalum**, Nutt. Leaflets obovate or narrower, toothed; flowers spicate
2. T. **Andersoni**, Gr. Densely silky: leaflets entire, acute: flowers umbellate.
3. T. **Lemmoni**, Watson. Leaflets coarsely serrate: flowers reflexed: ovules 2.
4. T. **Plummeræ**, Wat. Matted, hoary: leaflets 3 to 5, oblanceolate: ovary hairy.

§ 2. *Leaflets 3: heads not involucrate, terminal or apparently so, pedunculate: flowers sessile or nearly so (except No. 12): only No. 13 annual (its heads in one form sessile).*

5. T. **eriocephalum**, Nutt. Flowers in dense spikes, soon reflexed: ovary hairy.
6. T. **plumosum**, Dougl. Similar: flowers not reflexed; ovary smooth.
7. T. **longipes**, Nutt. Similar: ovoid heads smaller: nearly glabrous.
8. T. **altissimum**, Dougl. Leaflets very acute: 4 calyx teeth curved or twisted.
9. T. **Beckwithii**, Brewer. Leaflets broader: heads globose: calyx teeth straight.
10. T. **Kingii**, Watson. Leaflets acute: rachis produced: flowers rose-purple.
11. T. **Bolanderi**, Gr. Cæspitose the short stems decumbent: ovary smooth, 2-ovuled.
12. T. **Breweri**, Watson. Very slender, diffuse: flowers few, nearly white, pediceled.
13. T. **Macræi**, H. & A. Erect slender, 6 to 12 inches high: heads ovoid: peduncles long, or, in var. **dichotomum**, Brewer, short. (?)

§ 3. *Leaflets 3: heads small, not involucrate, pedunculate, axillary: flowers on short, at length reflexed pedicels: glabrous annuals: ovules 2.*

14.   **T.**   gracilentum, T. & G.   Flowers pale-rose to red-purple: calyx-teeth subulate.
15.   **T.**   ciliatum, Gr.   Similar: calyx teeth scarious margined, ciliolate.
16.   **T.**   bifidum, Gr.   Like 14 but smaller, the narrow leaflets bifid.

     § 4.   *Leaflets 3: heads subtended by an involucre: peduncles axillary: flowers in whorls, sessile or nearly so: annuals.*

         * *Involucre deeply many-cleft, laciniate: corolla not becoming inflated.*

17.   **T.**   involucratum, Willd.   Leaflets mostly oblanceolate, acute.
Var. heterodon, Watson.   Larger heads: leaflets broader: ovules same, mostly 5 or 6.
18.   **T.**   tridentatum, Lindl.   Slender and erect: leaflets very narrow: ovules 2.
Var. obtusiflorum, Watson.   Stout decumbent, glandular: leaflets broader.
Var. melanthum, Watson.   Smooth, slender: heads of dark purple flowers small.
19.   **T.**   pauoiflorum, Nutt.   Slender, weak: heads few-flowered: calyx teeth long.
20.   **T.**   monanthum, Gr.   Decumbent stems 1 to 4 inches long: heads 1 to 4-flowered.

   * * *Involucre light green, often whitish-scarious at base, not deeply lobed, broad as the the head, and saucer-shaped or cup-like: corolla not becoming inflated, or moderately so in No. 24.*

21.   **T.**   microcephalum, Pursh.   Soft hairy: involucre about 9-lobed, lobes entire.
22.   **T.**   microdon, H. & A.   Involucral lobes 3-toothed: calyx-teeth scarious-serrulate.
23.   **T.**   cyathiferum, Lindl.   Smooth: bristly-branched calyx - teeth equaling the corolla.   Eel Riv. and Sierra Val. to the Columbia Riv.
24.   **T.**   barbigerum, Torr.   Mostly less than a span high: calyx teeth bristly, long.
Var. Adrewsii, Gr.   Stouter, more hairy: plumose calyx teeth very long.

     * * * *Involucre rotate, lobes entire or wanting: corolla inflated in fruit.*

25.   **T.**   fucatum, Lindl.   Yellowish or white flowers often reddish tinged, large.
26.   **T.**   depauperatum, Desvaux.   Slender: flowers small: involucre often a ring.
Var. amplectans has a larger 4-5-parted involucre.   Heads in both forms small.

5.   MELILOTUS, Tournefort.

1.   **M.**   parviflora, Desf.   Flowers yellow a line long: spikes slender.   (Sweet Clover.)
2.   **M.**   officinalis, Willd.   Similar flowers 2 lines long on slender pedicels.
3.   **M.**   alba, Lam.   Flowers white.   All introduced from Eu.   The first common.

6.   MEDICAGO, Linnæus.

1.   **M.**   sativa, L.   Flowers blue-purple in close nearly capitate racemes.   (Alfalfa.)
2   **M.**   denticulata, Willd.   Flowers yellow: pod globose-coiled, prickly.   (Bur Clover.)
3.   **M.**   maculata, Willd.   Similar: leaflets with a dark spot.   All introduced from Eu.

7.   HOSACKIA, Douglas.

Flowers solitary or rarely 2 in the axils: no stipules.
     Peduncle bracteate or rarely naked.................................... **9 to 18**

Peduncle none or very short.. 14, 15
Flowers in pedunculate umbels or whorls.
  Peduncle with a compound or simple bract.   -†
    Bract below the top of the peduncle: stipules large................. 1,'2, 3
    Bract at the top of the peduncle.
      Stipulate leaves smooth................................ 4, 5
      Stipulate leaves pubescent...................... 1, 6, 7, 8
      Stipules none.   Pod with 5 or more seeds........... 8 to 12
                 Pod 1-2-seeded........... 17, 19, 22 to 25
  Peduncle not bracteate.   Stout, erect: pod more than 5-seeded................. 4
                 Slender, prostrate: pod 1-2-seeded............... 19, 20
Flowers in nearly or quite sessile umbellate clusters: pod 1-2-seeded.
      Nearly smooth, somewhat woody................... 16, 18
      Very pubescent (silky or tomentose)......... 21, 23, 24, 25

§ 1. *Pod linear, straight or nearly so, 5-20-seeded (2-4- seeded in 15) glabrous or nearly so (except in 10 and 16).*

* *Leaves with stipules, leaflets 5 to 20: umbels pedunculate: flowers 6 lines long or longer: keel obtuse: erect perennials.*

    † *Flowers dull-colored, yellowish and purple.*

1. **H. incana**, Torr. Low, stout, densely silky: peduncles about 6 lines long.
2. **H. stipularis**, Benth. Taller, villous, glandular: peduncles longer.
3. **H. crassifolia**, Benth. Erect, tall, nearly or quite glabrous: peduncles long.

    † † *Flowers rather showy, larger.*

4. **H. bicolor**, Dougl. Glabrous: flowers yellow with whitish wings.
5. **H. gracilis**, Benth. Similar: larger flowers with purplish wings.
6. **H. oblongifolia**, Benth. Flowers yellow and purple, standard orange.
Var. **angustifolia**, Watson. A span high: leaflets narrow: umbels 1-5-flowered.
7. **H. Torreyi**, Gr. Silky: standard yellow, wings and keel white.
8. **H. macrantha**, Greene. Stipules deciduous: petals yellow standard 6 lines broad.

* * *Stipules reduced to dark, often minute, glands (see No. 8): leaflets 3 to 9 (1 to 3 in No. 14): claws of petals not exserted.*

† *Peduncles long, 1-8-flowered: flowers exceeding 5 lines long: perennials: more or less appressed silky: leaflets obovate or narrower, rather acute.*

9. **H. grandiflora**, Benth. Flowers yellowish or greenish white, rarely purple.
10. **H. rigida**, Benth. Petioles short or none: flowers yellow, becoming brown.

† † *Peduncles 1-5-flowered, about equaling the leaves: flowers less than 6 lines long: yellow in 11, 12; pinkish in 13, 14.*

11. **H. maritima**, Nutt. Leaflets fleshy, 4 to 6 lines long, obovate or narrower.
12. **H. strigosa**, Nutt. Diffuse, strigose: leaflets linear, rarely ovate, small.
13. **H. parviflora**, Benth. Very slender: flowers 2 or 3 lines long, rarely yellow.
14. **H. Purshiana**, Benth. Widely branching, silky: flowering July to October.

† † † *Flowers nearly sessile and mostly solitary, not bracteate (see 12): leaves with a broad rachis which bears 3 to 5 leaflets at the end and one side.*

15 **H. subpinnata,** T. & G.   Much branched, usually decumbent or ascending and a few inches high: flowers yellow. Very common in Central Cal.

Var. major. Erect, corymbosely branched above, 6 inches to 3 ft. high, flowers pinkish. Northern Cal. to Washington.

16. **H. brachycarpa,** Benth.   Soft-hairy: flowers yellow: hairy pod 2-4-seeded.

§ 2. *Pod with a long slender incurved beak, 1-2-seeded: claw of the standard remote from the rest: umbels sessile or on short peduncles (except Nos. 18, 20): flowers less than 6 lines long: stipules minute dark colored glands: leaflets 3 to 7.*

* *Nearly glabrous: slender stems virgately branched: pod only slightly pubescent, 2-seeded.*

   † *Somewhat woody at the base: stems angled: leaflets mostly 3.*

17. **H. glabra,** Torr.   Stems very many erect or decumbent: leaves and fl's crowded.

18. **H. cytisoides,** Benth.   Similar: peduncles with a bract: calyx-teeth often recurved.

19. **H. juncea,** Benth.   Leaflets broader: some of the flowers pedunculate.

   † † *Not woody: stems terete: leaflets usually 5 to 7, and 2 or 3 lines long.*

20. **H. prostrata,** Nutt.   Leaflets obovate, acute: flowers 2 or 3 lines long.

21. **H. micrantha,** Nutt.   Flowers smaller: peduncle naked: style hairy.

* * *Very silky or tomentose; herbaceous stems terete: pod hairy: mostly 1-seeded: leaflets 5 to 7 (usually 3 in 22).*

22. **H. sericea,** Benth.   Densely white-silky: umbels loosely few-flowered.

23. **H. argophylla,** Gr.   Umbels 6-10-flowered: flowers orange or yellow. Sierras.

24. **H. decumbens,** Benth.   Villous and tomentose: stems diffuse: lvs and fls crowded.

25. **H. tomentosa,** H. & A.   Very tomentose, prostrate: flowers 3 or 4 lines long.

26. **H. Heermannii,** D. & H.   Less tomentose more branched: leaflets and fl's smaller.

8. PSORALEA, Linnæus.

Leaflets 3, orbicular on long petioles from creeping stem............................ 1
Leaflets 3, rombic-ovate to narrowly ovate: stems erect..................... 2, 3, 4
Leaflets 5, rarely 7: clustered stems very short................................. 5

1. P. **obicularis,** Lindl.   Peduncles a foot or two long. In wet ground. Cal.

2. P. **strobilina,** H. & A.   Stems 2-3 ft. high: peduncles shorter than the leaves. Cal.

3 P. **macrostachys,** DC.   Often 6 ft. high or more: peduncles exceeding the leaves.

4. P. **physodes,** Dougl.   Stems numerous, 1 or 2 ft. high: flowers greenish. Coast.

5. P. **Californica,** Watson.   Silky-gray: leaves exceeding the close racemes. Rare

9. AMORPHA, Linnæus.

1. A. **Californica,** Nutt.   Glandular, 3 to 10 ft. high: standard exceed by stamens.

10. GLYCYRRHIZA, Linnæus.

1. G. **lepidota,** Nutt. **var. glutinosa,** Watson.   Flowers white or pinkish, 6 lines long

11. ASTRAGALUS, Tournefort.

* *Annuals: pods 2-celled.*

Pods 1 or 2 lines long, 2-seeded, wrinkled: spikes short.......................... 1, 2
Pod linear, straight, 5 to many seeds: flowers capitate.......................... 3, 4
Pods 3 to 5 lines long: spikes of small white flowers very long...................... 5
Pods ovoid, long-beaked, gray-silky: flowers capitate, white....................... 6
  * * *Perennials: pods 1-celled, with thin walls, inflated, bladder-like.*
Pods on stipes equaling or little exceeding the calyx........................ 7, 8, 9
Pods on filiform, stipes much exceeding the calyx: stem erect............. 10, 11, 12
Pods sessile in the calyx, 1-2 inches long: many seeds...................... 13 to 17
                5 to 8 lines long: stems low: flowers 3 lines long...... 18, 19
* * * *Perennials: pods turgid, not bladder-like, coriaceous, densely long-woolly or downy,
incurved*... 20, 21, 22
  * * * * *Perennials: pods often turgid, not bladder-like, not long-hairy or woolly.*
Pods stipitate, 1-celled, sutures not inflexed............................ 23 to 26
                2-celled: cross section obcordately 2-lobed......................... 27
Pods not stipitate, 1 or 2 inches long, 1-celled............................. 28, 29
                2 or 3 lines long, 2-celled................................. 30
                2 lines long, hoary, cylindric-oval......................... 31
                3 lines long, 1-celled: leaflets spiny-tipped.................... 32

1. **A. didymocarpus,** H. & A. Calyx equaling the erect pod, black-hairy.
2. **A. nigrescens,** Nutt. Calyx ½ as long as the pendulous lightly wrinkled pods.
3. **A. tener,** Gr. Violet and white flowers: pods 5-7 lines long, drooping.
4. **A. Rattani,** Gr. Flowers larger, violet: pods slender, 1-1½ inches long, erect.
5. **A. Clevelandii,** Greene. Tall: leaflets ½-⅔ in. long, mucronate.
6. **A. Breweri,** Gr. Similar in habit to No. 4: pods 3-4 lines long, beaks longer.
7. **A. Hookerianus,** Dietr. Diffuse, silky, a span high: pod obovoid, obtuse.
8. **A. oxyphysus,** Gr. Erect, 2 or 3 ft. high, silky: pod slender-obovoid, acuminate.
9. **A. curtipes,** Gr. Lower, not silky: stipules united: pod ovoid or oval, acute.
10. **A. leucophyllus,** T. & G. Oval pod one-sided, filiform hairy stipe very long.
11. **A. leucopsis,** T. & G. Similar pod tapering into a smooth stipe half as long.
12. **A. trichopodus,** Gr. Pods smaller, 6 lines long or more, stipe 3 lines long.
13. **A. oocarpus,** Gr. Straggling stems 3-6 ft. long: green stipules mostly deflexed.
14. **A. Crotalariæ,** Gr. Scarious stipules distinct: ovoid pod 1-1½ inches long.
15. **A. Menziesii,** Gr. Similar: upper stipules united: pod larger, more bladdery.
16. **A. macrodon,** Gr. Like the preceding: flowers smaller: peduncles short.
17. **A. Douglasii,** Gr. Spike an inch long or less: pod ovoid 1½-2 inches long
18. **A. Hornii,** Gr. Pods in a dense head or short spike, ovoid, acuminate, hairy.
19. **A. Pulsiferæ,** Gr. White-hairy: pods few ovoid, curved, 3-8-seeded, hairy.
20 **A. Purshii,** Dougl. Tufted, silky: peduncles 5-6-flowered: pod ovoid.

21. A. Andersoni, Gr.   Densely white-hairy: leaflets 13-25 pairs: pods falcata.
22. A. Congdoni, Watson.   Less hairy: leaflets 8-10 pairs: pod narrower.
23. A. Gibbsii, Kellogg.   Soft-hairy: pod much curved, an inch or more long.
24. A. collinus, Dougl.   Hoary: pod slightly curved, erect, less than 1 inch long.
25. A. Californicus, Greene.   Stouter: pod straight, purple-bloched, 1½ inches long.
26. A. Antiselli, Gr.   Ashy-hairy; leaflets 21-29: straight pod, 8-9 lines long.
27. A. Bolanderi, Gr.   Scarious stipules united: pod ovoid, curved, veiny.
28. A. Webberi, Gr.   Silvery-silky: pods thick-walled, glabrous, sutures prominent.
29. A. pychnostachyus, Gr.   Stout, hoary: pods reticulated, thin-walled, acute.
30. A. Lemmoni, Gr.   Slender, procumbent, green: leaflets 9-11, mucronate.
31. A. Austinæ, Gr.   Tufted, silvery-silky: flowers in a close head, whitish.
32. A. Kentrophyta, Gr.   Flowers 1-3 on very short peduncles, 2 lines long.

12. VICIA, Tournefort.

Perennials: peduncles 4-18-flowered... 1, 2
Annuals: peduncles short, 1-2-flowered... 3, 4
1. V. gigantea, Hooker.   Stout, 5-10 ft. high: petals dull-purplish.
2. V. Americana, Muhl.   Glabrous, 1-4 ft. high: leaflets 8-16, variable.
Var. truncata, Brewer.   Leaflets truncate or toothed at apex, somewhat hairy.
Var. linearis, Watson.   Leaflets linear: mostly low and slender.
3. V. exigua, Nutt.   Mostly low: leaflets about 8: flowers 3 lines long, purplish.
4. V. sativa, Linn.   Stouter: flowers nearly sessile, 1 inch long, violet.

13. LATHYRUS, Linnæus.

Leaves with tendrils: racemes several—many-flowered.
Pod not on a stipe.   Stipules large: glabrous................................. 1, 2, 3
                        Stipules narrow: more or less pubescent.................... 7, 8
Pod on a short stipe.   Stems stout, tall.................................... 4, 5, 6
Leaves without tendrils, or rarely with them: pods on short stipes......... 9, 10, 11
1. L. maritimus, Bigelow.   Leaflets 3-5 pairs, close flowers purple.
2. L. polyphyllus, Nutt.   Similar: leaflets 6-10 pairs, thin, not sessile.
3. L. sulphureus, Brewer.   Flowers sulphur or dull yellow, 5-7 lines long.
4. L. Nuttallii, Watson.   Loosely woolly-hairy: petals red-purple, 6-8 lines long.
5. L. Californicus, Watson.   Stem winged: leaflets soft-pubescent: petals 7-9 lines
      long, yellowish or pinkish.   This and next under L. venosús, in Bot. Cal., etc.
6. L. Bolanderi, Watson.   Stems wingless: glabrous: flowers purple.
7. L. vestitus, Nutt.   Slender: stems wingless: flowers pale rose or violet.
8. L. palustris, L.   Leaflets 2-4 pairs, linear: flowers few, purplish, small.
9. L. litoralis, Endl.   Densely silky: a small terminal leaflet: pod hairy.
10. L. Nevadensis, Watson.   Slender: standard purplish; wings and keel yellowish.

11. **L. Torreyi, Gr.** Acute leaflets 6 lines long: purplish flowers solitary.

14. CERCIS, Linnæus.

C. occidentalis, Torr. Small standard enclosed by the wings: pods red.

ROSACEÆ.

* Shrubs or Trees.

a. Flowers white, 3 lines or more across: carpels 1-5, distinct.

Carpel 1, becoming a drupe (like a cherry or plum)............................. **1**

Carpels 5 (or often wanting), stamens 15: racemes drooping: cherry-like............. **2**

Carpels 2-5, becoming inflated, usually reddening: stamens 20 or more............... **7**

Carpels 5, hairy: leaves bipinnate, leaflets minute: panicles leafy..................... **8**

Carpel 1, becoming an akene: low shrub: leaves tripinnate........................... **9**

b. Flowers white, 2 lines broad or less, in dense panicles: carpels 1 to 12.

Stamens 20 or more: flowers in plumose panicles: leaves large, lobed................. **6**

Stamens 10-15: evergreen leaves 2-4 lines long: panicles erect........................**18**

c. Flowers rose-color or pale purple, small: carpels 5, distinct..................... **3**

d. Flowers 3 lines or more across: carpels 2-5, enclosed by the fleshy calyx-tube forming an inferior ovary (partly inferior in 24).

Evergreen leaves serrate: carpels 2: stamens 10: berries scarlet....................**24**

Deciduous leaves simple: flowers corymbose.

    Fruit red or yellow: no spines or thorns...............................**25**

    Fruit black or purple: spinose...**26**

Deciduous leaves simple: flowers racemose: petals oblong..........................**27**

e. Flowers 6 lines broad or more: stamens and carpels numerous: fruit like a blackberry or raspberry...**10**

f. Flowers solitary, axillary, small: petals none: calyx white, the limb deciduous: carpel 1 (rarely 2), long plumose tailed. **11**

g. Flowers rose-colored an inch or more across: stamens many: ovary apparently inferior: stems slender, prickly: leaves pinnate.....................................**23**

* Herbs.

a. Flowers 6 lines broad or more: akenes forming a berry............................**10**

b. Flowers very small (except 1 sp. in 5), white: calyx lobes 5 (no intermediate lobes or bractlets): stamens 20 or more: carpels 3-10.

Leaves very large, tripinnate: spikes in large panicles............................. **4**

Leaves twice or thrice 3-cleft: raceme short......... .... **5**

Leaves entire, rosulate, silky: scape low.................................. 3d sp. in **3**

Leaves 5-7-lobed with small basal leaflets: petals 2-3 lines long............ 4th sp. in **8**

h

a. *Flowers yellow, white or purple*: *calyx appendaged between the lobes, or apparently un-
        equally 10-lobed.*
    Stamens 20 or more: carpels very numerous: receptacle conical to clavate
        Akenes with hooked or plumose tails...12
        Akenes seed-like on a juicy receptacle: leaves 3-foliolate....................13
        Akenes seed-like on a dry receptacle. (Try No. 16 and 17.)...................14
    Stamens 20 or less: carpels few or many on a dry receptacle.
        Stamens 10 (or 20 and corolla pink): corolla white: receptacle nearly naked....16
        Stamens 5 to 20; filaments slender: carpels few or 1: receptacle hairy.........17
        Stamens 5: carpels 5 to 10: leaflets 3, cuneate, 3-5-toothed..................15
d. *Flowers small: carpels 1 to 3 becoming akenes enclosed by the firm calyx-tube forming a
        apparently inferior ovary.*
    Leaves pinnate flowers in heads or spikes.
        Calyx with barbed prickles: petals none: anthers purple.......................21
        Calyx with a margin of hooked prickles: petals yellow.........................20
        Calyx 4-angled, naked, limb petaloid: petals none.............................22
    Leaves palmately lobed: greenish apetalous flowers axillary.......................19

1. PRUNUS, Tournefort.

Flowers white: scraggy or spiny: leaves ovate, an inch long or less................ **1, 3**
            branches slender: leaves 1 to 4 inches long....................... **2, 4**
            low: leaves spatulate, entire, 6 lines long, sessile..................... **7**
            evergreen leaves, shining, prickly toothed, broad................. ...... **5**
Flowers rose-color: low, spiny: leaves oblanceolate, 6 to 12 lines long............... **6**
    [The first species is a small plum, the others scarcely edible cherries.]
1. P. subcordata, Benth. Bark ashy gray: flowers in small lateral clusters.
2. P. emarginata, Walpers. Bark chestnut brown: corymbs 6-12-flowered.
Var. mollis, Brewer. Taller, becoming 25 ft. high, woolly. Oregon.
3. P. Fremonti, Walp. Flowers solitary or few together: ovary densely hairy.
4. P. demissa, Walp. Dense racemes 3-4 inches long, erect: leaves large.
5. P. ilicifolia, Walp. Racemes small, axillary: fruit ripening in November.
6. P. Andersoni, Gr. Low, diffuse: leaves oblanceolate, acute: fruit velvety.
7. P. fasciculata, Gr. Similar: slender petals recurved: stamens 10-15:

2. NUTTALLIA, Gray.

1. N. cerasiformis, T. & G. Diœcious: 1 to 4 of the carpels maturing drupes.

3. SPIRÆA, Linnæus.

* *Shrubs with rose-colored or purplish flowers: carpels 5.*
1. S. betulæfolia: Pallas. Pale purple flowers in corymbs. Alpine.

2. S. **Douglasii**, Hooker. Darker flowers in dense panicles. Wet places.
  * * Herbaceous with a woody base: flowers white.
3. S. **cæspitosa**, Nutt. Tufted: flowers in dense spikes on leafy scapes.
4. S. **occidentalis**, Watson Simple glabrous stems 2-6 ft. high: panicle cymose.

4. ARUNCUS. Linnæus.

1. A. **sylvester**, Kost. Smooth, 3-5 ft. high: flowers diœcious: stamens exserted.

5. ERIOGYNIA, Hooker.

1. E. **pectinata**, Hook. Cespitose, creeping; branches erect: stamens included.

6. HOLODISCUS, Maximowicz.

1. H. **discolor**, Max. Flowers mostly dull white or light buff: carpels hairy.

7. PHYSOCARPUS, Maximowicz.

1. P. **opulifolia**, Max. Bark shreddy: leaves 3-lobed: corymbs 2 inches broad.

8. CHAMÆBATIARIA, Maximowicz.

1. C. **Millefolium**, Max. Leaves narrowly lanceolate, 1-3 inches long.

9. CHAMÆBATIA, Bentham.

1. C. **foliolosa**, Benth. Strong scented, viscid: leaves ovate to oblong.

10. RUBUS. Linnæus.

Stems woody: leaves simple, palmately lobed: no prickles............. 1
Stems woody: leaves mostly 3-foliolate: more or less prickly................. 2, 3, 4
Stems herbaceous, trailing, not prickly: carpels few............................ 5, 6
1. R. **Nutkanus**, Mocino. Large leaves: large rose-like flowers.
2. R. **leucodermis**, Dougl. Leaves white below, veins pricky. (Raspberry.)
3. R. **spectabilis**, Pursh. Flowers large, red-purple: fruit yellow or crimson.
4. R. **ursinus**, C. & S. Stems weak, often long-trailing: very prickly. (Blackberry.)
5. R **pedatus**, Smith. Leaves 3-foliolate or nearly 5-foliolate: fruit red.
6. R. **lasiococcus**, Gr. Stouter: leaves mostly 3-5-lobed: fruit tomentose.

11. CERCOCARPUS, HBK.

1. C. **parvifolius**, Nutt. Evergreen: leaves veiny, serrate above: wood hard.
2. C. **ledifolius**, Nutt. Leaves narrow, entire margins revolute. (Mt. Mahogony.)

12. GEUM, Linnæus.

1. G. macrophyllum, Wild.   Flowers yellow: style jointed: akene-tails hooked.
2. G. triflorum, Pursh.   Flowers purplish: styles plumose: akene-tails feathery.

13. FRAGARIA, Tournefort.

* *Akenes deeply pitted in the depressed-globose fruit.*

1. F. Chilensis, Ehrh,   Leaves thick, dark green, shining: flowers large.   Coast.
2. F. Virginiana, Ehrh.   Similar: flowers smaller: fruit darker.
* * *Akenes on the surface of the ovoid fruit.*
3. F. Californica, C. & S.   Light green thin leaves: petioles not silky.
4. F. vesca, L.   Similar: larger.   Perhaps No. 3 is only a variety of this.

14. POTENTILLA, Linnæus.

* *Style attached at or below the middle of the ovary.*

A foot or two high: leaflets 5-9, coarsely serrate: petals yellow or white.............. 1
Creeping: leaves and peduncles radical: leaflets 7 to many: petals yellow............. 2
Stems stout, rooting at the joints: flowers dark lurid-purple........................ 3
Shrubby leaflets entire, silky, margins revolute................................ 4
* * *Style attached at or near the top of the ovary: stamens 20.*
Alpine or subalpine (altitude 7,000 ft. or more) leaflets an inch long or less.
            Densely white tomentose: leaflets 7 to 13: carpels stipitate................. 5
            Silky-villous: leaflets closely pinnate or palmate....................... 8, 9
            Glabrous: leaflets 3, broadly cuneiform, 7-9-toothed...................... 11
In the mountains but mostly lower than 7,000 ft.
  .   Leaflets 5-15, deeply serrate or pinnatifid.............................. 6, 7
            Leaflets 3, toothed above... 10, 11
1. P. glandulosa, Lindley.   Petals usually shorter than the calyx.
Var. Nevadensis, Watson.   More slender: stamens sometimes only 20.
2. P. Anserina, L.   Long runners: leaflets with smaller ones between.   Wet places.
3. P. palustris, Scop.   Many fibrous roots: leaves palmate: leaflets serrate. Swamps.
    P. fruticosa, L.   Much branched: stamens 30: carpels very villous.
5. P. Breweri, Watson.   Leaflets nearly equal, 3-6 lines long: petals large.
6. P. Plattensis, Nutt.   Slender stems 3-12 inches long: leaflets pinnatifid.
7. P. gracilis, Dougl.   Taller, more hairy: leaflets and flowers larger.
Var. rigida, Watson.   Tall and stout, not tomentose.   The common form.
8. P. dissecta, Pursh.   Leaflets pinnatifid or coarsely serrate: tufted-hairy.
9. P. Wheeleri, Watson.   Leaflets cuneate, 3-5-toothed, 6 lines long or less.
10. P. Grayi, Watson.   Low stems 3-6-flowered: leaflets 5-7-toothed.
11. P. gelida, C. A. Meyer.   Leaflets larger, broader, and one nearly sessile.

15. SIBALDIA, Linnæus.

1. **S. procumbens**, L. Stems creeping: calyx lobes exceeding the minute petals.

16. HORKELIA, C. & S.

Styles thickened at the base: leaflets incised 6-12 lines long......................... 1
Calyx-bracts nearly as broad as the lobes: leaflets 3-8 lines long..................... 2
Calyx-bracts much narrower than the lobes.
    Leaflets deeply incised or lobed............................... 3, 4, 5, 6
    Leaflets few-toothed at the truncate apex............................... 7
    Leaflets bifid, 2 or 3 lines long, silky................................... 8
1. **H. fusca**, Lindl. Cymes dense: petals 2 lines long, sepals longer.
2. **H. Californica**, C. & S. Glandular: sepals and petals 3-6 lines long.
Var. sericea, Gr. Stouter: leaflets larger: canescent with silky hairs.
3. **H. congesta**, Hook. Sparsely stiff-hairy: leaflets 6-9 lines long.
4. **H. tenuiloba**, Gr. White-hairy: leaflets 8-12 pairs, 2-3 lines long.
5. **H. Bolanderi**, Gr. Densely hoary, pubescent, tufted, 3-4 inches high.
6. **H. purpurascens**, Watson. Purplish calyx 3-4 lines long: petals rose-color.
7. **H. tridentata**, Torr. Silky: leaflets 2-5 pairs, mostly 3-toothed at apex.
8. **H. sericata**, Watson. Tufted: some stem leaflets entire: petals notched.

17. IVESIA, Torrey & Gray.

Flowers in rather close panicled cymes: stems mostly leafy. 1, 2, 3
Flowers yellow, in cymes on nearly naked stems. Alpine...................... 4, 5
Flowers white, in diffuse panicles upon leafy stems................................ 6
1. **I. Pickeringii**, Torr. Densely white-silky: petals yellowish, spatulate.
2. **I. unguiculata**, Gr. Similar: petals white, clawed, orbicular: carpels 5-8.
3. **I. Webberi**, Gr. Low, loosely villous: petals yellow: stamens 5-10.
4. **I. Gordoni**, T. & G. Viscid: 3-10 inches high: stem leaves pinnatifid.
Var. pygmæa, Watson. An inch or two high: stamens sometimes 10.
Var. lycopsoides, Watson. Nearly glabrous: leaflets thick, rounded, imbricated.
5. **I. Muirii**, Gr. Densely silky, an inch high: leaves terete: carpels 2.
6. **I. santalinoides**, Gr. Stems 6-18 inches high: leaves silky: terete: carpel 1.

18. ADENOSTOMA, H. & A.

1. **A. fasciculatum**, H. & A. Bark becoming shreddy: leaves subulate, acute.
2. **A. sparsifolium**, Torr. Resinous: leaves scattered, obtuse. San Diego.

19. ALCHEMILLA, Tournefort.

1. **A. arvensis**, Scopoli. An obscure under-herb: stipules enclosing the flowers.

20. AGRIMONIA, Tournefort.

ı. **A. Eupitoria, L.** Hairy: 2-4 ft. high: racemes spicate: akene subglobose.

21. ACÆNA, Linnæus.

ı. **A. trifida, R. & P.** Leaves crowded at base: green flowers in terminal spike.

22. POTERIUM, Linnæus.

1. **P. officinale, B. & H.** Flowers deep purple or red in a short spike.
2. **P. annuum,** Nutt. Smaller: leaflets pinnatifid: flowers greenish.

23. ROSA, Tournefort.

1. **R. Nutkana,** Presl. Spines stout: stipules broad; flowers 2 or 3 inches broad.
2. **R. pisocarpa,** Gr. Globose fruit smaller, with a neck.
3. **R. Californica,** C. & S. Often tall: fruit ovoid, with a neck.
4. **R. spithamea,** Watson. A span high or less: globose fruit glandular-prickly.
5. **R. gymnocarpa,** Nutt. Slender: calyx-lobes deciduous, leaving fruit naked.

24. HETEROMELES, J. Rœmer.

ı. **H. arbutifolia,** Rœm. Panicles terminal: fruit ripe in December. (*Toyon.*)

25. PIRUS, Linnæus.

1. **P. rivularis,** Dougl. Leaves simple, woolly: fruit ovoid.
2. **P. sambucifolia,** C. & S. Leaves pinnate: fruit globose, red.

26. CRATÆGUS, Linnæus.

1. **C. rivularis,** Nutt. Leaves ovate, serrate, rarely lobed. (*Hawthorn.*)
2. **C. Douglasii,** Lindl. A large tree: leaves often incised: fruit 6 lines thick.

27. AMELANCHIER, Medicus.

ı. **A. alnifolia,** Nutt. Flowers in short racemes: petals not ovate. (*Shad-berry.*)

CALYCANTHACEÆ.

1. CALYCANTHUS, Linnæus.

ı. **C. occidentalis, H. & A.** Inner sepals and outer petals lurid purple or red, ᴀ inch or more long, slender, leathery: inner petals shorter, incurved.

SAXIFRAGACEÆ.

§ 1. Herbs with leaves alternate or all radical (except No. 2 in 7): styles 2 or 3 (sessile stigmas 3 or 4 in 11): carpels united or rarely distinct, the tips divergent (flattened-obcordate in 12).

Stamens apparently many in clusters, only 5 perfect............................. 11
Stamens 8-10: petals none: flowers minute, axillary, solitary..................... 12
Stamens 10 (rarely more or less in No. 1): flowers in terminal racemose or cymose clusters.

 Petals broad, entire: ovary 2-celled or carpels distinct....................... 1
 Petals pinnatifid, 3-7 lobed or entire: ovary 1-celled: racemes simple.......... 7
 Petals entire, very slender: styles slender; carpels unequal................... 8
Stamens 5: petioles expanded stipule-like, or bristly at base.

 Petals deciduous, entire, broad: radical leaves 3-parted.............. No. 2 in 1
  leaves 3-9-lobed or cleft: ovary inferior. 2
 Petals deciduous, often crenate, white: seeds winged...................... 5
  pinnatifid or 3-cleft; lobes filiform: ovary globular........... 9
  entire or wanting, small: ovary 1-celled....................10
 Petals persistent, entire or 3-lobed, spatulate, violet: ovary inferior............ 3
  entire, slender, purple: ovary superior..................... 4
Stamens 3: petals entire, filiform, recurved persistent.............................. 6
§ 2. *Shrubs with large white flowers or woody-based herbs: leaves opposite.*
 Shrubs: stamens 20 or more: petals 4 or 5: stigmas distinct...................13
  petals 5 to 7: stigmas united...................14
 Herbaceous: branches terminated by capitate clusters of small flowers.........15
§ 3. *Slender shrubs: leaves alternate: flowers mostly in drooping clusters: ovary inferior, globose: calyx-lobes larger than the erect petals, usually petaloid.*...................16

1. SAXIFRAGA, Linnæus.

 * *Stemless, or rarely a leaf or two on the scape below.*
Leaves large, peltate, cupped in the center: flowers pink............................. 1
Leaves an inch or less broad, short petioled: scape 2 to 4 inches high................ 2
Leaves rounded-cordate, long petioled: filaments broadest above: calyx soon reflexed 8, 9
Leaves not cordate, contracted at base into a margined petiole or nearly sessile.

 Calyx-lobes erect or spreading.................................... 3, 5
 Calyx-lobes reflexed in fruit or sooner..:..................... 4, 6, 7
 * * *Stems leafy, tufted (except No. 12): petioles short.*
 Leaves small, evergreen, entire, crowded. Alpine......................... 10
 Leaves like strawberry leaflets: flowers greenish........................... 11
 Leaves few on the stem: stamens 5....................................... 12
1. S. peltata, Torr. Scape stout, 1-3 ft. high: carpels distinct. Streams.
2. S. Parryi, Torr. Calyx and white petals brown or purple-veined.

3. **S. Virginiensis,** Michx.   Petals obovate, twice the length of the calyx.
4. **S. reflexa,** Hook.   Calyx reflexed: filaments often thick above.
5. **S. nivalis,** L.   Flowers fewer, more crowded; petals narrow, small.
6. **S. integrifolia,** Hook.   Scape 1-3 ft. high, viscid: seeds large.
7. **S. bryophora,** Gr.   Leaves slender, entire: many pedicels bulb-bearing.
8. **S. Mertensiana,** Bong.   Leaves many-lobed: pedicels often bulbiferous.
9. **S. punctata,** L.   Leaves coarsely toothed: panicle not bulbiferous.
10. **S. Tolmiei,** T. & G.   Peduncles 2 inches long: carpels often 3 or 4.
11. **S. fragarioides,** Greene.   Woody caudex branched: petals persistent.
12. **S. ranunculifolia,** Hook.   Slender, a foot high or less: flowers corymbose.

2. BOYKINIA, Nuttall.

1. **B. occidentalis,** T. & G.   Leaves thin, 1-3 inches broad, incisely toothed.
2. **B. major,** Gr.   Stouter, larger: leaves 4-8 inches broad, 5-9-cleft.

3. SUKSDORFIA, Gray.

1. **S. violacea,** Gr.   Weak, viscid: flower parts rarely in 7's.   Or. Wash.

4. BOLANDRA, Gray.

1. **B. Californica,** Gr.   Stems slender: petals dull purple.   Yosemite.
2. **B. Oregana,** Watson.   Stouter: petals deep purple: pedicels reflexed in fruit.

5. SULIVANTIA, Torrey & Gray.

1. **S. Oregana,** Watson.   Scape nearly leafless: leaves an inch or less broad.

6. TOLMIEA, Torrey & Gray.

1. **T. Menziesii,** T. & G.   Calyx gibbous, finally splitting down one side.

7. TELLIMA, Robt. Brown.

Petals laciniately pinnatifid, reflexed, rose-color or greenish........................ 1
Petals entire, spatulate-obovate, white or pinkish: calyx-base acute, adnate.......... 2
Petals entire or with small side-teeth, obovate or oval, white: calyx-base broad...... 3
Petals 3-lobed, nearly cuneate, white or pinkish: calyx-base broad: styles smooth..... 4
                                         calyx-base turbinate, styles rough... 5
Petals deeply 3-cleft, pink or white: ovary half inferior: bulblet-bearing............. 6
Petals palmately 3-7-parted, mostly pink: ovary nearly free: bulblet-bearing......... 7
1. **T. grandiflora,** R. Br.   Stout, 1-3 ft. high: calyx inflated.   Monterey to Alaska.
2. **T. Cymbalaria,** Walp.   Stem filiform, usually a pair of leaves.   S. Cal.
3. **T. Bolanderi,** Bol.   Stems often branching: styles smooth.   Cent. Cal.

4. T. heterophylla, H. & A. Similar, very slender: petals acute. Cent. Cal.
5. T. affinis, Bolander. Stouter: calyx-tube rough, partly adnate. Cal.
6. T. parviflora, Hook. Calyx obconical: ovary half inferior. Or. Wash.
7. T. tenella, Walp. Slender, 2-9 in. high: ovary nearly free. N. E. Cal.

8. TIARELLA, Linnæus.

1. T. unifoliata, Hook. Leaves 3-5-lobed, crenately toothed. S. F. Bay, N.
2. T. trifoliata, L. Most of the leaves 3-foliolate. Or. to Alaska.

9. MITELLA, Tournefort.

1. M. Breweri, Gr. Leaves 2-3 in. broad: scape naked: petals ¼ in. long. S.N. Mts.
2. M. trifida, Grah. Petals smaller, 3-5-parted. In shade. Coast Ranges.
3. M. caulescens, Nutt. Stouter: 1 or more leaves on scape. Klamath River, N.

10. HEUCHERA, Linnæus.

1. H. glabra, Willd. Leaves acutely lobed, incised: panicle loose. Or., N.
2. H. rubescens, Torr. Leaves crenately lobed ½-1 in. broad. S. N. Mts., N. & E.
3. H. micrantha, Dougl. Leaves obtusely lobed, crenate, 2-4 in. broad.
4. H. pilosissima, F. & M. Very villous, viscid: calyx-base rounded. Cal. Coast.
5. H. cylindrica, Dougl. Flowers nearly spicate, ¼-½ in. long: petals minute.

11. PARNASSIA, Tournefort.

1. P. palustris, L. Leaves ovate to ovate-cordate ¼-1 in. long: petals ¼-½ in. long.
Var. Californica, Gr. Larger in every way. In wet places, mountains.
2. P. fimbriata, Banks. Leaves reniform to cordate: appendages few cr a scale.

12. CHRYSOSPLENIUM, Linnæus.

1. C. glechomæfolium, Nutt. Decumbent in wet places: leaves ¼-½ in. long.

13. PHILADELPHUS, Linnæus.

1. P. Lewisii, Pursh. Spreading, 3-5 ft. high: stigmas narrow. S. N. Mts.
2. P. Gordonianus, Lindl. Larger in every way: leaves mostly toothed. Coast.

14. CARPENTERIA, Torrey.

1. C. Californica, Torr. Leaves narrowly lanceolate. King's River, Cal.

15. WHIPPLEA, Torrey.

1. W. modesta, Torr. Stems slender, spreading: calyx and corolla white.

16. RIBES, Linnæus.

§ 1. More or less thorny and prickly: leaves 3-5-lobed, parted or divided: peduncles 1-5-flowered (5-9-flowered in No. 10), (*Gooseberries.*)

Calyx bright red: fruit bristly or prickly.................................... 1, 2, 4
Calyx yellow: leaves very small; fruit smooth.................................. 5, 7
Calyx white or pinkish, lobes erect: ovary white-villous; fruit velvety.............. 6
Calyx greenish, villous: stamens short: fruit spiny-prickly......................... 3
Calyx greenish or dull-purplish: ovary and fruit smooth...................... 8, 9
Calyx greenish white, limb saucer-shaped, 3 lines broad: stamens short........... 10
§ 2. Thornless, no prickles: stamens short: berry not prickly. (*Currants.*)
Calyx greenish white, rotate above the ovary: berry ¼-⅓ in. long, black............. 11
Calyx waxy-white, greenish or pinkish; tube cylindrical, ¼-½ in. long.............. 12
Calyx dull white, greenish or purplish; tube cylindrical-campanulate............... 13
Calyx rose-red to nearly white; tube short, broad: racemes dense: fruit dry.......... 14
Calyx golden yellow, salverform; tube ¼-⅓ in: long: spicy-fragrant................. 15

1. R. speciosum, Pursh. Tall: Flowers nearly an inch long, parts often in 4's.
2. R. Menziesii, Pursh. Very thorny: anthers sharp-sagittate.
3. R. ambiguum, Watson. Glandular, villous: white petals nearly as long as the stamens.
4. R. Lobbii, Gr. Flowers 9 lines long: anthers oval: very obtuse, white.
5. R. leptanthum, Gr. Rigid, much branched: style undivided, smooth.
6. R. velutinum, Greene. Rigid recurved branches: stout thorns solitary.
7. R. quercetorum, Greene. Calyx-lobes ciliate, reflexed, bright yellow.
8. R. divaricatum, Dougl. Stems widely spreading: stamens exserted.
9. R. oxycanthoides, L. Similar: flowers smaller; stamens little exserted.
10. R. lacustre, Poiret. var. molle, Gr. Low: leaves downy: berry light red.
11. R. bracteosum, Dougl. Leaves 5-7-cleft, lobes acute, 5-9 in. wide.
12. R. cereum, Dougl. Leaves 3-lobed, an inch broad or less, crenate.
13. R. viscosissimum, Pursh. Viscid: leaves moderately lobed.
14. R. sanguineum, Pursh. Leaves obtusely 3-5-lobed: petals often white: variable.
15. R. aureum, Pursh. Glabrous or nearly so: racemes 5-10-flowered.

CRASSULACEÆ.

Obscure under herbs: minute greenish flowers in the axils of minute leaves.......... 1
Very fleshy herbs: leaves entire (serrate in 1st sp. No. 2): petals distinct............. 2
        petals more or less united... 3

1. TILLÆA, Linnæus.

1. T. minima, Miers. Flowers clustered in the axils: leaves ovate, a line long.

2. **T. angustifolia,** Nutt. Stems rooting, 1 in. long: leaves 1-2 lines long.
Var. Bolanderi, Watson. Stems 3-4 in. high: flower-parts in 3's or 4's.
3   **T. peduncularis,** Smith. Pedicels 4-6 lines long in fruit: carpels purplish.

2. SEDUM, Linnæus.

Flowers diœcious, deep purple, parts mostly in 4's.................................... 1
Leaves narrowed toward the base, obtuse.................................... 2, 3, 4
Leaves broadest near the base, acute.................................... 5, 6, 7
1. **S. Rhodiola,** DC. Stems simple: leaves acute, rarely entire. Alpine.
2. **S. spathulifolium,** Hook. Glaucous: petals yellow, 3 lines long, acute.
3. **S. Oreganum,** Nutt. Not glaucous: petals pale rose, much exceeding the stamens.
4. **S. obtusatum,** Gr. Similar to 2 & 3: flowers pedicelled; petals pale yellow, broader.
5. **S. variegatum,** Watson. Radical leaves slender: petals yellow, often purple-veined.
6. **S. radiatum,** Watson. Carpels broad, the beaks horizontal, star-like: petals yel.
7. **S. pumilum,** Benth. Annual; 1-3 inches high: leaves 1-2 lines long: carpels 1-seeded.

3. COTYLEDON, Linnæus.

Leaves cylindrical and somewhat 3-sided.................................... 1, 2
Leaves flattened: carpels nearly distinct, erect.
       Petals united to the middle, red.............................. 3
       Petals united ½ the length, pale yellow, 4 lines long............. 4
       Petals united only at the base, yellow or orange.
              Leaves glaucous, white dusty or mealy... 5, 6, 7, 8
              Leaves not glaucous or mealy, reddish........ 9, 10
1. **C. edulis,** Brewer. Petals widely spreading, yellowish. San Diego.
2. **C. viscida,** Watson. Leaves numerous, very viscid: corolla reddish. S. Cal.
3. **C. pulverulenta,** B. & H. Densely white-dusty: leaves 2 or 3 inches broad.
4. **C. Oregonensis,** Watson. Leaves spatulate, obtuse: racemes axillary.
5. **C. lanceolata,** B. & H. Petals orange, mid-vein glaucous: calyx-base broad.
6. **C. farinosa,** B. & H. Usually densely mealy: petals lemon yellow.
7. **C. cæspitosa,** Haworth. Sepals ovate, 2 lines long or less: petals yellow.
8. **C. laxa,** B. & H. Petals orange-yellow, keeled, prominent mid-vein glaucous.
9. **C. Palmeri,** Watson. Flowering stem red: petals pale yellow. S. Cal. coast.
10. **C. Lingula,** Watson. Leaves an inch broad, 2 or 3 inches long, acute.

DROSERACEÆ.

1. DROSERA, Linnæus.

1. **D. rotundifolia,** L. Leaf-blade rounded, 2-6 lines broad: petals 2 lines long.

2. D. Anglica, Hudson. Leaf-blade cuneate: petals much exceeding the sepals.

LYTHRACEÆ.

Calyx slightly 4-angled, short: petals none: capsule globular...... 1
Petals 4: capsule striate under microscope, splitting into 3 or 4 valves................ 2
Calyx cylindrical, striate, 4 to 7 teeth with smaller ones between: petals 4 to 7 (usually
    6): stamens as many or twice as many: capsule 2-celled......................... 3

1. AMMANIA, Linnæus.

1. A. latifolia, L. Stems square: leaves opposite, eared at base, slender.

2. ROTALA, Kœhne.

1. R. ramosior, Kœh. Capsule not bursting irregularly.

3. LYTHRUM, Linnæus.

1. L. hyssopifolia, L. Petals very small: stamens usually 4 to 6, included.
2. L. album, HBK. Calyx 3 or 4 lines long.
3. L. Californicum, Watson. Calyx 6 to 9 lines long: rose-purple petals longer.

HALORAGEÆ.

Leaves linear or broader in whorls of 4 to 12: flowers sessile, axillary................ 1
Leaves in whorls of 3 or 4, mostly pinnate, segments filiform: flowers axillary........ 2

1. HIPPURIS, Linnæus.

1. H. vulgaris, L. Style filiform, stamen 1: calyx globular, limb entire. In ponds.

2. MYRIOPHYLLUM, Linnæus.

1. M. spicatum; L. Stamens 8: flowers spicate: petals ovate, greenish.
2. M. hippurioides, Nutt. Stamens 4: petals white, obovate: toothed bracts 3 to 5
    lines long. Both species aquatic.

ONAGRACEÆ.

Calyx divided to the ovary, lobes persistent: aquatic or creeping marsh herbs: solitary flowers
in axils of entire leaves.
Petals 4 to 6, yellow: stamens 8 to 12: leaves alternate.......................... 1

Petals none or 4, reddish: stamens 4: leaves opposite............................... 2
* * *Calyx deciduous above the ovary: parts of the flowers in 4's.*
† *Anthers versatile (attached near the middle to the filament).*
a. Seeds with a tuft of silky hairs, stamens 8: lower leaves often opposite.
Calyx-tube long-funnelform above the ovary, scarlet: petals 2-lobed.................. 3
Calyx-limb 4-parted: anthers elliptical or roundish................................ 4
b. Seeds not tufted with hairs: leaves all alternate: calyx-lobes reflexed.
Calyx divided nearly or quite to the ovary: anthers of two forms (or only 4).
   Leaves entire: small or minute flowers: capsules not an inch long............. 5
   Leaves pinnatifidly lobed: capsules 3 or 4 inches long............ 6
Calyx forming a cup or tube (often long and slender) above the ovary.............. 7
†† *Anthers erect, attached at or near the base to the filaments; those opposite the petals shorter or sterile, rarely wanting: calyx-lobes reflexed, except in No. 11: annuals.*
Calyx-tube obconical above the ovary: petals not long-clawed...................... 8
                                      petals long-clawed.......................... 9
Calyx-tube filiform above the ovary: petals long-clawed, lobed......................10
Calyx-lobes erect: petals not clawed, 2-lobed: stigma lobes short...................11
Calyx-lobes spreading: petals clawed, entire: stigma discoid, entire.................12
* * * Minute white flowers in bractless racemes, the parts in 2's...................13

1. JUSSIÆA, Linnæus.

1. **J. repens, L.** Stems a foot or more long, rooting at base: style stout, hairy.
Var. Californica, Watson. Smaller flowers 6 to 8 lines broad: style slender, glabrous.

2. LUDWIGIA, Linnæus.

1. **L. palustris, Ellis.** Smooth, creeping or floating: flowers sessile: capsule small.

3. ZAUSCHNERIA, Presl.

1. **Z. Californica, Presl.** Villous or tomentose: calyx 10-16 lines long above ovary.

4. EPILOBIUM, Linnæus.

* *Flowers large: stamens and style declined: stigma lobes finally spreading or recurved: perennial.*
Stem erect, simple: petals clawed, obovate, lilac-purple......................... 1, 2
Stem decumbent, 5 to 3 inches long: leaves opposite, glaucous..................... 3
Flowers yellow.. 4
   * * *Flowers small, parts all erect: stigma club-shaped or cylindrical.*
Perennial: petioles present but short: in wet places: not alpine................. 5, 6
Perennial: stems with 2 pubescent lines: leaves mostly opposite: alpine........... 7, 8
Perennial: leaves sessile: hoary or glaucous: leaves mostly opposite: not alpine. 9, 10, 11

ONAGRACEÆ.

Annual: stems round: leaves mostly alternate: petals obcordate............ **12, 13, 14**
1. E. spicatum, Lam. Simple stem often 5 or 6 ft. high: flowers spicate.
2. E. latifolium, Linn. Shorter, less erect, often branching: style glabrous.
3. E. obcordatum, Gr. Petals obcordate, spreading, rose-color. Alpine.
4. E. luteum, Pursh. Oregon to Alaska.
5. E. Watsoni, Barbey. Hoary-pubescent, branching: petals narrow, obcordate.
6. E. coloratum, Muhl. Erect, branched, puberulent: leaves mostly opposite.
7. E. alpinum, L. Creeping, 2 to 6 inches high: flower-buds ovoid: anthers globose.
8. E. origanifolium, Lam. Taller: large petals obcordate, purple to white.
9. E. Franciscanum, Barb. Stems 2 to 4 ridged: petals purple, emarginate.
10. E. brevistylum, Barb. Similar: petals smaller, obcordate, pinkish. S. N. Mts.
11. E. glaberimum, Barb. Glaucous: leaves connected at base: petals notched.
12. E. paniculatum, Nutt. Often tall, very slender: leaves slender, often fascicled.
13. E. jocundum, Gr. Similar: panicles crowded: petals larger, 6 lines long, deep purple.
14. E. minutum, Lindl. Less than a ft. high: petals minute.

5. GAYOPHYTUM, A. Jussieu.

1. G. ramosissimum, T. & G. Diffuse, 6 to 18 inches high: capsule 3-5-seeded.
2. G. racemosum, T. & G. Less branched, more flowers, capsule 8 to 10 lines long.
3. G. pumilum, Watson. Smaller: capsule 6 lines long, the many seeds oblique.
4. G. diffusum, T. & G. Flowers 1½ to 3 lines broad, usually pink: capsules on pedicels.

6. EULOBUS, Nuttall.

1. E. Californicus, Nutt. Stout, 1 to 3 ft. high: petals 4 or 5 lines broad, yellow.

7. ŒNOTHERA, Linnæus.

§ *Calyx-tube much prolonged beyond the ovary: stigma-lobes slender.*
Tall: flowers yellow, erect in bud: seeds in 2 rows in each cell...................... 1
Stems white: flowers white or purplish, 1½ or 2 inches broad: buds nodding 2, 3, 4
Stemless: calyx-tube 2 to 4 inches long: capsule wing-angled....................:..... 5
§ 2. *Calyx-tube filiform, stem-like above the ovary: stigma capitate: flowers yellow, sessile on the top of a rootstock in the axils of radical leaves.*
Nearly glabrous: leaves lanceolate to ovate: perennial.......................... 6, 7
Canescently villous: leaves linear: capsule 4-winged: annual........................ 8
§ 3. *Calyx-tube obconic or short-funnelform: stigma capitate: capsules narrow, sessile or nearly so, often contorted.*
Flowers axillary, yellow: capsule sharply 4-angled, usually contorted........ 9, 10, 11
Flowers axillary, small, yellow: capsule obtusely angled, curved..... .... 12, 13

Flowers small, in a nodding spike, white or rose-color: capsule contorted....... **14, 15**
1. Œ. biennis, L. Stout, 1 to 5 ft. high: calyx-tube 1 to 2½ inches long.
Var. grandiflora, Lindl.   Capsule hirsute: petals as long as calyx-tube.
Var. hirsutissimus, Gr. Similar: ovary more hirsute. The common form.
2. Œ. albicaulis, Nutt. Erect, ½ to 4 ft. high: large leaves pinnatifid.
3. Œ. Californica, Watson. Hoary, decumbent: ovary and calyx villous.
4. Œ. trichocalyx, Nutt. Stouter, more erect: capsule thicker at base.
5. Œ. triloba, Nutt. Nearly glabrous: petals ½ to 1½ inches long.
6. Œ heterantha, Nutt. Petals 3 to 6 lines long: capsules ovoid-oblong.
Var. taraxacifolia, Watson. Leaves lyrately pinnatifid: Sierra Nevada.
7. Œ. ovata, Nutt. Similar: capsule with obtuse angles. Coast Valleys.
8. Œ. graciliflora, H. & A. Petals obcordate, turning greenish, 3 to 5 lines long.
9. Œ. cheiranthifolia, Hornem. Prostrate or ascending: leaves thick.
Var. suffruticosa, Watson. Woody at base, leafy. Both on sand near the sea.
10. Œ. bistorta, Nutt. Similar: petals 4 to 9 lines long, spot at base.
11. Œ. micrantha, Hornem. Flowers smaller: leaves thin, crimped.
12. Œ. dentata, Cav. Diffuse, very slender: leaves linear: capsule very slender.
13. Œ. strigulosa, T. & G. Similar: petals smaller, 1 or 2 lines long, reddening.
14. Œ. alyssoides, H. & A. Slender, canescent: calyx-lobes and petals 2 or 3 lines long.
15. Œ. gauræflora, T. & G. Glabrous: bark loose, white: capsule tapering upward.

8. GODETIA, Spach.

* *Flowers in strict mostly close spikes: stems leafy: capsule ovoid to oblong.*
† *Capsule not ribbed: seeds in 2 rows in the cell: stems simple or few branched.*
Calyx-tube broadly obconical, 4 to 6 lines long: petals 1 or 2 inches long............. 1
Calyx-tube 2 or 3 lines long, deep-purple.. 2
† † *Capsule with at least alternate sides 2-ribbed: seeds in 1 row in each cell: stems often much branched: canescently puberulent.*
Capsule 5 to 8 lines long tapering from the base.................................. 3
Capsule 3 to 6 lines long, oblong, short-hairy.................................... 4
* * *Flowers mostly scattered in a simple spike or raceme and nodding in the bud: capsule linear; seeds in 1 row: stems slender: leaves few.*
Capsules sessile, more or less distinctly ribbed................................ 5 to 9
Capsules on pedicels, not ribbed: stigmas mostly yellow................... 10 to 13
1. G. grandiflora, Lindl. Stout: spike leafy: petals often with a spot. N. W. Cal.
2. G. purpurea, Watson. Ovary densely villous: style short: stigma-lobes purple.
3. G. lepida, Lindl. Stem shining white: petals rose-color with a spot above.
Var. parviflora, Watson. Petals smaller, ¼-⅔ in. long: leaves ½-1 in. long.
Var. Arnottii, Watson. Nearly glabrous: leaves longer, acute: capsule glabrous.
4. G. albescens, Lindl. Flowers small, in many spikelets: petals purple-blue.
5. G. Williamsoni, Watson. Petals yellowish at base, purple spot in center. Cal

6. **G. quadrivulnera**, Spach. Petals purple, $\frac{1}{4}$-$\frac{1}{2}$ in. long: stigma-lobes purple. Coast.
7. **G. tenella**, Watson. Petals similar: style short: capsule scarcely ribbed. Coast.
8. **G. viminea**, Spach. Similar to 7: petals like 6 (or all purple) but larger.
9. **G. Romanzovii**, Spach. Ovary silky: 4 of the anthers nearly sessile: style short.
10. **G. amœna**, Lilja. Petals (and purple anthers) often villous, rose-color to white.
11. **G. Bottæ**, Spach. Petals light purple: stigma yellow or purple. Monterey, S.
12. **G. epilobioides**, Watson. Petals purple to white, $\frac{1}{4}$-$\frac{1}{2}$ in. long: calyx-tube short.
13. **G. hispidula**, Watson. Hispid; often 1-flowered: 8-10 in. high. Cent. Cal.
14. **G. biloba**, Watson. A span to 4 ft. high: petals 2-lobed: rose-purple. Cent. Cal.

9. CLARKIA, Pursh.

1. **C. pulchella**, Pursh. Petals 3-lobed; claw 2-toothed: only 4 perfect stamens. Or.
2. **C. Xantiana**, Gr. Petals 2-lobed, with a tooth between; claw short. S. Cal.
3. **C. elegans**, Dougl. Petals entire; claw long, slender, naked: capsule hairy. Cal.
4. **C. rhomboidea**, Dougl. Petals entire; claw short, broad, often toothed.

10. EUCHARIDIUM, Fischer & Meyer.

1. **E. concinnum**, F. & M. Petals 3-lobed: filaments slender: capsule 6 to 9 lines long.
2. **E. Breweri**, Gr. Petals deeply 2-lobed, with a tooth between: calyx-tube longer.

11. BOISDUVALIA, Spach.

Canescently pubescent and more or less villous...................................... 1, 2
Glabrous or slightly pubescent: loosely spicate.................................... 3, 4
1. **B. densiflora**, Watson. Leafy spikes dense: petals 3-6 lines long.
2. **B. Torreyi**, Watson. Floral leaves like the others: petals 1 or 2 lines long. Or. S.
3. **B. glabella**, Walp. Similar: petals smaller: seeds a line long. Or. & Wash.
4. **B. cleistogama**, Curran. Capsules curved outward: seeds numerous, minute.

12. HETEROGAURA, Rothrock.

1. **H. Californica**, Rothr. Petals spatulate, 2 lines long: fruit obovoid, nut-like. Cal.

14. CIRCÆA, Linnæus.

1. **C. Pacifica**, A. & M. Leaves ovate to cordate, opposite: ovary ovoid, hairy. Woods.

LOASACEÆ.

1. MENTZELIA, Linnæus.

Leaves mostly sinuate-toothed: petals 1 to 3 lines long.......................... 1, 2
Leaves mostly pinnatifid, lanceolate: petals 3 to 8 lines long..................... 3, 4

Leaves pectinately pinnatifid to sinuate-toothed: petals 1 to 2½ inches long........ **5, 6**
1. **M.** dispersa, Watson. Only upper leaves ovate: seeds ½ line long.
2. **M.** micrantha, T. & G. Leaves ovate, 1 inch long or less: seeds a line long.
3. **M.** congesta, T. & G. Bracts membranous at base: petals 3 to 6 lines long.
4. **M.** gracilenta, T. & G. Petals obovate to oblanceolate, 4 to 8 lines long.
5. **M.** Lindleyi, T. & G. Leaves ovate or narrower: petals ovate. Cent. Cal.
6. **M.** lævicaulis, T. & G. Stout: leaves 2 to 8 inches long: petals 2 to 2½ inches long.

CUCURBITACEÆ.

Flowers all solitary, large, yellow: seeds flat.. 1
Flowers small, white; the sterile racemose: seeds turgid............................ 2

1. CUCURBITA, Linnæus.

1. **C.** perennis, Gr. Leaves 6-12 in. long: flowers 3 and 4 in. long, fragrant.
2. **C.** palmata, Watson. Leaves 2-3 in. long, 5-cleft: calyx-tube an inch long.
3. **C.** Californica, Torr. Flowers an inch long or more; calyx 4 or 5 lines long.

2. MEGARRHIZA, Torrey.

1. **M.** Californica, Torr. Fruit globose or ovoid: seeds 4, 8-12 lines long.
2. **M.** macrocarpa. Fruit ovoid oblong, 14-seeded. Santa Barbara, south.
3. **M.** Marah, Watson. Fruit 4 in. long: seeds suborbicular, flattened.
4. **M.** Oregana, Torr. Fruit scarcely or not at all armed with spines. Or.-Wash.
5. **M.** muricata, Watson. Fruit globose, an inch long, 2-seeded, mostly naked.
   Perhaps this genus should be united with *Echinocystis*, which has flat seeds.
   The second species is described by E. L. Greene under the name *Echinocystis macrocarpa*.

DATISCACEÆ.

1. DATISCA, Linnæus.

1. **D.** glomerata, B. & H. Diœcious or perfect flowers in leafy spikes, greenish.

CACTACEÆ.

Oval or cylindrical plants, covered with spine-bearing tubercles..................... 1
Depressed-globose plants with tuberculate ribs and woolly at top: spines stout, ringed 2
Cylindrical ribbed stems branching, 2 to 4 ft. high: spines numerous................ 3
Stems made up of flattened or cylindrical joints: spines barbed..................... 4
9

130   FICOIDEÆ.

1. MAMILLARIA, Haworth.
1. M. Goodridgii, Schear. Petals about 8, ovate, awned, dull yellowish. S. Cal.

2. ECHINOCACTUS, Link & Otto.
1. E. viridescens, Nutt. Sepals and petals numerous, greenish: fruit scaly. S. Cal.

3. CEREUS, Haworth.
1. C. Emoryi, Engelm. Flowers greenish yellow: fruit subglobose, spiny. S. Cal.

4. OPUNTIA, Tournefort.
Joints much flattened, obovate: fruit juicy. 1, 2
Joints cylindrical: fruit green, fleshy: flowers red................................. 3
1. O. Engelmanni, Salm. Flowers yellow, 3 inches long. Santa Barbara, S.
2. O. Ficus-Indica, Mill. Fruit delicious. The Prickly Pear.
3. O. prolifera, Engelm. Tree-like with spiny tubercles. San Diego.

FICOIDEÆ.

Ovary inferior: petals and stamens numerous: very fleshy........................ 1
Ovary superior: petals none: leaves opposite or whorled.
    Calyx-lobes 5, petaloid: stamens many: succulent.... 2
    Sepals 5, greenish: stamens 3 to 10: styles 3: not succulent................... 3

1. MESEMBRYANTHEMUM, Linnæus.
1. M. æquilaterale, Haworth. Leaves equally 3-sided, very thick, opposite.
2. M. coccineum, Haw. Leaves semi-cylindrical, a line broad.
3. M. crystallinum, L. Leaves flat, covered with glistening papillæ.

2. SESUVIUM, Linnæus.
1. S. Portulacastrum, L. Calyx 3 to 5 lines long, more or less purplish.

3. MOLLUGO, L.
1. M. verticillata, L. Slender: leaves spatulate: flowers fascicled, axillary.

UMBELLIFERÆ.

Herbs with usually hollow stems, sheathing petioles and small flowers in simple or compound umbels; the inferior ovary surmounted by a disk which bears 5 petals and 5 stamens: styles 2. The plants of this order are not here described.

CORNACEÆ.

Flowers in cymes or heads: petals 4: style 1: ovary 2-celled........................... 1
Flowers diœcious, in drooping blue-gray catkins: petals none: styles 2.............. 2

1. CORNUS, Linnæus.

Flowers in a head with involucre of 4 to 6 large white petaloid bracts........... 1, 2
Flowers yellowish in sessile umbels appearing before the leaves: bracts 4........... 3
Flowers white or cream-color in many-flowered cymes........................ 4 to 7
1. C. Canadensis, L. Herbaceous, 3 to 8 inches high: leaves in a whorl at top.
2. C. Nuttallii, Audubon. A tree: involucre often tinged with red. (Dogwood.)
3. C. sessilis, Torr. Bark green: leaves pale and silky beneath: pedicels silky.
4. C. Californica, C. A. Meyer. Branches purplish: leaves ovate: cyme round-topp'd.
5. C. pubescens, Nutt. Similar: leaves rarely ovate: cyme larger: fruit white.
6. C. glabrata, Benth. Bark gray: leaves acute at each end: cymes small, flat.
7. C. Torreyi, Watson. Leaves ovate or narrower, acute: cyme loose: fruit white.

2. GARRYA, Douglas.

Leaves undulate, base obtuse, tomentose beneath 1
Leaves not undulate, acute at each end.. 2, 3
1. G. elliptica, Dougl. Sterile aments 2 to 5 inches long, often clustered.
2. G. Fremonti, Torr. Petioles 4 to 6 lines long: aments solitary, 2 or 3 inches long.
3. G. buxifolia, Gr. Smaller, 2 to 5 ft. high: leaves 1 to 1½ inches long, silky beneath.

DIVISION II.—GAMOPETALÆ.

CAPRIFOLIACEÆ.

§ 1. *Corolla regular, rotate: style short; stigmas 3 to 5: flowers in cymes.*
Shrub or tree: young stems thick, pith large: leaves pinnate....................... 1
Slender shrub: leaves simple: fruit a drupe: seeds flat......................... 2
§ 2. *Corolla tubular and irregular or campanulate: stigma 1, capitate.*
A creeping herb with pendulous flowers in pairs................................ 3

Slender shrubs with small pinkish regular flowers: berries white......... 4
Shrubby climbers or erect shrubs with irregular tubular flowers..................... 5

1. SAMBUCUS, Tournefort.

Cymes round-topped; pith of year-old shoots yellow-brown..................... 1, 2
Cymes flat-topped, 1-sided: pith white: berries black........................... 3, 4
1. S. racemosa, L. Cymes ovoid or oblong: flowers dull white: berries scarlet.
2. S. melanocarpa, Gr. Cymes broader: flowers white: fruit black, no bloom.
3. S. glauca, Nutt. Leaves smooth: fruit black with a white bloom. (*Elder-berry*.
4. S. Mexicana, Presl. Leaves and long shoots hairy: fruit black, no bloom.

2. VIBURNUM, Linnæus.

Drupes light red, globose, acid: leaves all or some of them lobed................. 1, 2
Drupes bluish-black, flattened, elliptical: leaves not lobed......................... 3
1. V. opulus, L. Cymes on several-leaved branches. Or. & Wash., E. & N.
2. V. pauciflorum, Pylaie. Cymes on short 2-leaved branches. Wash., N. & E.
3 V. ellipticum, Hooker. Leaves 3-5-ribbed: corolla 4 or 5 lines broad. N. Cal., N.

3. LINNÆA, Gronovius.

1. L. borealis, Gronov. Corolla funnelform, 4 or 5 lines long: stamens 4.

4. SYMPHORICARPOS, Dillenius.

Corolla broadly campanulate, 2 or 3 lines long............................'........... 1, 2
Corolla narrowly campanulate, 3 to 6 lines long............................... 3, 4
1. S. racemosus, Michx. Smooth: corolla hairy, narrow at base.
2. S. mollis, Nutt. Mostly soft-hairy, diffuse: corolla broad at base.
3. S. rotundifolius, Gr. Leaves orbicular to elliptical, 6 to 9 lines long.
4. S. oreophylus, Gr. Corolla 4 to 6 lines long, scarcely hairy: nutlets sharp.

5. LONICERA, Linnæus.

Erect shrubs: flowers in axillary pairs on a single peduncle.
    Ovaries ⅔ or wholly united to form a single berry: bracts subulate.......... 1, 2
    Ovaries distinct or nearly so: corolla saccate at base, yellowish............. 3, 4
Woody climbers: flowers sessile, clustered: upper leaves often united............. 5, 6
1. L. cærulea, L. Only 1 or 2 ft. high: leaves pale: corolla yellowish or purplish.
2. L. conjugialis, Kellogg. Corolla dull purple, bilabiate, 4 or 5 lines long.
3. L. Utahensis, Watson, Bracts small: berries red: peduncles short.
4. L. involucrata, Banks. Bracts large, becoming red: berries purple black.
5. L. ciliosa, Poir. Corolla an inch long, yellow to crimson-scarlet.
6. L. hispidula, Dougl. Corolla 6 lines long, the lobes half as long: variable.

RUBIACEÆ.

A shrub with opposite or whorled leaves: flowers in globular heads................. **1**
Herb with opposite leaves: flower parts in 4's (rarely 3's or 5's): fruit bristly.......... **2**
Herbs with whorled leaves: stems square: flowers 3-4-merous: fruit biglobular....... **3**

1. CEPHALANTHUS, Linnæus.

1. C. occidentalis, L. Corolla narrow funnelform, white, 4-lobed. (*Button-bush.*)

2. KELLOGGIA, Torrey.

1. K. galioides, Torr. Corolla funnelform, 3 or 4 lines long, pinkish or white.

3. GALIUM, Linnæus.

Fruit dry: leaves all in 4's, or the upper in pairs...................... 2, 3, 4, 8, 9
leaves mostly in 6's (some in 4's, 5's or 8's).................... 1, 5, 6, 7
Fruit juicy: perennials with leaves in 4's................................... 10 to 14
1. G. Aparine, L. Retrorsely hispid: leaves in 6's and 8's: fruit erect.
2. G. bifolium, Watson. Smooth: alternate leaves shorter: peduncles solitary.
3. G. Kamtschaticum, Steller. Leaves orbicular to oblong-ovate, 3-nerved.
4. G. boreale, L. Leaves narrow, 3-nerved: flowers white in terminal panicles.
5. G. trifidum, L. Leaves slender, obtuse, 4 to 7 lines long: flower parts often in 3's.
6. G. asperrimum, Gr. Leaves lanceolate, 6 to 12 lines long, cymes dichotomous.
7. G. triflorum, Michx. Sweet scented: corolla greenish or yellowish: cymes 3-rayed.
8. G. angustifolium, Nutt. Smooth, woody at base, rigid: fruit long-bristly.
9. G. multiflorum, Kellogg. Tufted, a foot high or less: leaves ovate.
10. G. pubens, Gr. Grayish, much branched: leaves broad, 6 lines long or less.
11. G. Californicum, H. & A. Similar: leaves hispid-ciliate. Coast Range.
12. G. Nuttallii, Gr. Tall, mostly smooth: leaves small, oval or narrower.
13. G. Bolanderi, Gr. Mostly smooth: corolla dull purple: berry white.
14. G. Andrewsii, Gr. Matted tufts 2 to 4 inches high, leaves crowded, narrow, shining, sharp.

VALERIANACEÆ.

Calyx-limb of plume-like lobes, inrolled until fruiting: leaves lobed or parted........ 1
Calyx-limb none: flowers in dense terminal clusters: leaves simple.................. 2

1. VALERIANA, Tournefort.

1. V. sylvatica, Banks. Stem leaves 3-11-foliolate: corolla 2 or 3 lines long.
2. V. Sitchensis, Bong. More robust: stem leaves 3-5-foliolate: corolla larger.

2. VALERIANELLA, Tournefort.

1. **V. macrocera**, Gr.   Corolla 1 or 2 lines long, nearly regular, white or pinkish.
2. **V. congesta**, Lindl   Stouter: corolla mostly 3 or 4 lines long, bilabiate limb.
3. **V. anomala**, Gr.   Freely branching: corolla a line long, spurless.
4. **V. aphanoptera**, Gr.   Slender: corolla a line long, bilabiate, spur short.
5. **V. samolifolia**, Gr.   Similar: fruit wingless, buckwheat-like.

DIPSACACEÆ.

1. DIPSACUS, Tournefort.

1. **D. fullonum**, L.   Stiff leaves united in pairs: fruit oval, scales hooked.   Nat.

COMPOSITÆ.

Sunflowers, marigolds, thistles and dandelions are types of the conspicuous plants in this order. It would be difficult for the beginner to determine the species in this order; hence it is omitted.

LOBELIACEÆ.

Ovary nearly superior: anthers distinct: branches zigzag: leaves minute............... 1
Ovary inferior: anthers united: flowers blue or red.
    Corolla red, an inch long: adnate calyx-tube hemispherical.................... 2
    Corolla blue, rarely purple, often with white or yellow on lower lip.
       Ovary top-shaped: corolla-tube 6 to 9 lines long, hairy inside............. 3
       Ovary obconical to club-shaped: peduncles long.......... 4
       Ovary slender, stalk-like, sessile often twisted......................... 5

1. NEMACLADUS, Nuttall.

1. **N. ramosissimus**, Nutt.   Corolla a line long: unequal calyx-lobes, exceeding capsule.
2. **N. longiflorus**, Gr.   Corolla 3 lines long: equal calyx-lobes shorter than capsule.

2. LOBELIA, Linnæus.

1. **L. splendens**, Willd.   Simple stem 2 or 3 ft. high, ending in naked raceme

3. PALMERELLA, Gray.

1. P. debilis, Gr. Stems very leafy, 1 or 2 ft. high, ending in leafy-bracted racme.

4. LAURENTIA, Micheli.

1. L. carnosula, Benth. Rooting in mud, 1 to 5 inches high: leaves entire.

5. DOWNINGIA, Torrey.

1. D. elegans, Torr. Often 9 to 12 inches high: leaves slender: corolla blue with white and yellow spot on lower lip like the following:
2. D. pulchella, Torr. Lower corolla lip broader than long.
3. D. bicornuta, Gr. Corolla lip with a pair of hollow appendages at base.
4. D. concolor, Greene. Slender, diffuse: corolla blue throughout.

CAMPANULACEÆ

Capsule club-shaped, crowned with the rigid calyx-lobes, opening on top.............. 1
Capsule oblong, opening by 2 or 3 holes in the sides: seeds flattened.................. 2
Capsule short, opening as in No. 2: flowers all with corolla: calyx-lobes slender....... 3
Capsule obpyramidal, bursting indefinitely: calyx-lobes ovate, toothed.... 4

1. GITHOPSIS, Nuttall.

1. G. specularioides, Nutt. Leaves small, coarsely toothed: flowers all alike.

2. SPECULARIA, Heister.

1. S. biflora, Gr. Leaves ovate to lanceolate: lower flowers apetalous, sepals 3 or 4.
2. S. perfoliata, A. DC. Stouter: leaves round, cordate-clasping: lower flowers similar.

3. CAMPANULA, Tournefort.

Annual: flowers erect; calyx-lobes connivent about the style in fruit................ 1
Perennials: calyx-lobes not connivent in fruit: corolla deeply lobed.
    Style not longer than the corolla........................... .... 2, 3, 4
    Style filiform, exceeding the corolla: leaves sharply serrate......... 4, 5, 6
1. C. exigua, Rattan. Branching and flowering from base, 2 to 8 inches high.
2. C. scabrella, Engelm. Whitened with short hairs, flowers erect, 5-6 lines long.
3. C. rotundifolia, L. Stem leaves linear: corolla bright blue, 6 to 12 lines long.
4. C. linnæifolia, Gr. Leaves broad, obtuse, crenately serrate: corolla light blue.
5. C. Scouleri, Hooker. Leaves ovate to lanceolate, short petioled: pedicels long.
6. C. prenanthoides, Durand. Leaves mostly sessile: flowers often clustered: pedicels short.

4. HETEROCODON, Nuttall.

L. H. rariflorum, Nutt. Stems filiform: leaves orbicular, toothed, small.

ERICACEÆ.

Suborder I. VACCINIEÆ.

Shrubs (some low and herbaceous): ovary inferior becoming an edible berry.... 1

Suborder II. ERICINEÆ.

Shrubs or trees: calyx free, usually small: corolla gamopetalous (except 11, 12).

* *Fruit berry-like or fleshy: flowers drooping: corolla ovoid to campanulate with small lobes: stamens 8 or 10 included: bark shedding from at least the branches: leaves evergreen, coriaceous.*

Tree: flowers in large panicles: orange-red berries many seeded. 2
Shrubs: flowers in small racemes: fleshy fruit, 1-10 seeded..... 3
Shrubs, low or prostrate: flowers axillary: berries black or red..................... 4

** *Fruit, a dry, many-seeded capsule: flowers nodding: anthers awn-tipped.*

Shrub, 3 or 4 ft. high: oblong leaves 1 to 3 inches long............................. 5
Shrub, a foot high or less: small scale-like leaves in 4 ranks........................ 6

* * * *Fruit a dry capsule, splitting between the cells: anthers not awned.*
† *Corolla gamopetalous.*

Low Alpine evergreen; leaves revolute: flowers umbellate or corymbose:
  Leaves linear, crowded corolla not pouched... 7
  Leaves oblong, opposite; corolla 10-ribbed, from 10 depressed pouches.......... 8
Not alpine: leaves crowded at the ends of branches, entire.
  Corolla usually 4-toothed. ovoid to cylindrical, dull purple................... 9
  Corolla usually 5-lobed, limb spreading, white to rose...................... 10

†† *Corolla polypetalous or nearly so.*

Flowers in corymbs or umbels, erect, white, cherry-like.... 11
Flowers solitary, nodding, reddish... 12

Suborder III. PYROLEÆ.

Perennials, herbaceous or slightly woody with smooth evergreen leaves (except one species in No. 15): flowers nodding, polypetalous; petals broad: ovary superior: stamens 10: anthers in bud extrorse, at length by inversion introrse with 2-horned base above.

Flowers umbellate or solitary on a leafy woody stem............................. 13
Flowers solitary on a short scape: petals spreading.............................. 14
Flowers in a raceme on a scape; petals concave, incurved...... 15

Suborder IV. MONOTROPEÆ.

Herbs, parasitic upon roots: stems juicy, scaly-bracted, not green.

Stem striped, red or purple and white: sepals and bracts white...................... 16
Stem brown-red or purplish-red, clammy, hairy.................................. 17
Stem very thick; entire plant bright red.. 18
Stem white, tawny or reddish, fleshy; 19 and 20 polypetalous.
    Sepals 2 to 5, bract-like: petals 3 to 6, concave at base: style tubular........ 19
    Sepals and petals 4 or 5 each, lacerate-fringed, flat.......................... 20
    Sepals 2 or 4, petals united; filaments and style hairy....................... 21

1. VACCINIUM, Linnæus.

* *Corolla ovoid or globose, 4-5-toothed: filaments smooth; anthers 2-awned on the back included: leaves deciduous.*
Flowers often 2 to 4 together; corolla usually 4-toothed, leaves entire............. 1, 2
Flowers solitary, axillary: corolla usually 5-toothed: calyx not deeply lobed.
    Usually less than a foot high; leaves serrate......................... 3, 4
    Usually several (1 to 12) ft. high; branches spreading.............. 5, 6, 7
** *Corolla obovoid or campanulate, 5-toothed: leaves evergreen....................... 8*
*** *Corolla deeply 4-parted, lobes reflexed, pale rose-color: leaves evergreen........... 9*
1. V. uliginosum, From a span to 3 or 4 ft. high: leaves thick and veiny.
2. V. occidentale, Gr. Leaves thinner, less veiny: flowers mostly solitary.
3. V. cæspitosum, Michx. Branches not angled: berries blue. Very variable.
4. V. Myrtillus, L. var. microphyllum, Hooker. Branches sharply angled.
5. V. myrtilloides, Hooker. Branchlets slightly angled: leaves serrulate, veiny.
6. V. ovalifolium, Smith. Smooth, 4 to 12 ft high; branchlets angled.
7. V. parvifolium, Smith. Smooth; branchlets green, jointed, sharply angled.
8. V. ovatum, Pursh. Rigid; leaves ovate or narrower, serrate: flowers clustered.
9. V. oxycoccus, L. var. intermedium, Gr. Trailing, slender: flowers umbellate.

2. ARBUTUS, Tournefort.

1. A. Menziesii, Pursh. Leaves 3 to 5 inches long; corolla white, broad-ovoid.

3. ARCTOSTAPHYLOS, Adanson.

a. Seeds not united or easily separable.
    Low or creeping, rising only a foot or two: flowers 1 or 2 lines long.
      Trailing or creeping, green, no bristly hairs, ovary and fruit glabrous.. 1, 2
      Erect: leaves mostly not an inch long: flowers more numerous...... 3, 4, 5
    Erect, 3-20 ft. high: flowers 3-4 lines long: fruit 4-5 lines thick.......... 6, 7, 8
b. Seeds united into a solid woody or bony stone.......................... 9 to 12
1. A. Uva-ursi, Spreng. Leaves oblong-spatulate, retuse, tapering to petiole.

2. **A. Nevadensis**, Gr. Leaves obovate or narrower, cuspidate-mucronate, obtuse at base.
3. **A. pumila**, Nutt. Tomentulose, pale leaves oblong-obovate obtuse or retuse.
4. **A. Hookeri**, Don. Diffuse: leaves green, ovate or oval, cuspidate or acuminata.
5. **A. nummularia**, Gr. Very leafy: leaves mostly broadly oval, ends rounded.
6. **A. Andersoni**, Gr. Leaves thin, bright green, base sagittate or cordate.
7. **A. tomentosa**, Dougl. Branchlets bristly: leaves pale, ovate or narrower.
8. **A. pungens**, HBK. Leaves rigid, oblong-lanceolate to round-ovate, entire.
    Var. **platyphylla**, Gr. Leaves paler, broader, 1 or 2 inches long; not cuspidata.
9. **A. glauca**, Lindl. Larger (8 to 24 ft. high): fruit larger: glabrous branchlets.
10. **A. bicolor**, Gr. Leaves tomentose beneath: flowers rose-color 3 or 4 lines long.
11. **A. Clevelandii**, Gr. More hairy; leaves narrower, sessile, acuminate.
12. **A. polifolia**, HBK. Leaves linear-lanceolate: fruit rough, purple.

4. GAULTHERIA, Linnæus.

Flowers in slender but stiff, often branching, bracteate, viscid racemes............... 1
Flowers axillary, solitary; filaments glabrous; anthers not awned................. 2, 3
1. G. **Shallon**, Pursh. Spreading, 1 to 4 ft. high; leaves 2 to 4 inches long; serrulate.
2. G. **Myrsinites**, Hooker. Spreading in tufts: leaves oval or orbicular ½ inch long.
3. G. **ovatifolia**, Gr. Larger: leaves broadly ovate to subcordate. Or. N.

5. LUCOTHOE, Don.

1. L. **Daviasa**. Torr. Flowers in terminal, often clustered racemes, white, S. N. Mts

6. CASSIOPE, Don.

1. C. **Mertensiana**, Don. Leaves keeled, not furrowed on back, 1½-2 lines long.
2. C. **tetragona**, Don. Leaves thick, deeply furrowed on back, often pubescent.
3. C. **lycopodioides**, Don. Stems creeping filiform: leaves barely a line long.

7. BRYANTHUS, Steller.

1. B. **Breweri**, Gr. · Corolla rose-purple, 5-cleft to the middle, 4-5 lines broad.
2. B. **empetriformis**, Gr. Corolla smaller slightly lobed: stamens included.

8. KALMIA, Linnæus.

1. K. **glauca**, Ait. Leaves glaucous, white beneath: flowers saucer-shaped. Alpina.

9. MENZIESIA, Smith.

1. M. **glabella**, Gr. Leaves obovate, usually obtuse: filaments ciliate below.
2. M. **ferruginea**, Sm. Leaves oblong or broadly oblanceolate, acute, rusty-hairy.

10. RHODODENDRON, Linnæus.

Deciduous: flowers from lateral buds, nodding; corolla nearly rotate.................. 1
 flowers from terminal buds; tube funnel-form; limb spreading............. 2
Evergreen: many-flowered corymbs terminal: corolla campanulate, lobes broad....... 3
1. R. albiflorum, Hooker. Low: corolla white, 5-cleft: stamens included.
2. R. occidentale, Gr. Taller: corolla white, viscid; stamens exserted.
3. R. Californicum, Hooker. Leaves 3 to 6 inches long: corolla rose-purple.

11. LEDUM, Linnæus.

1. L. latifolium, Ait. Leaves rusty-tomentose below, margins strongly revolute.
2. L. glandulosum, Nutt. Leaves whitish beneath, resinous, scarcely revolute.

12. CLADOTHAMNUS, Bongard.

1. C. pyrolæflorus, Bong. Tall, slender, smooth: sepals equaling the petals.

13. CHIMAPHILA, Pursh.

1. C. Menziesii, Spreng. Leaves often mottled above: peduncle 1-3-flowered.
2. C. umbellata, Nutt. Taller (1 or 2 ft. high) leaves not spotted: flowers 4 to 8.

14. MONESES, Salisbury.

1. M. uniflora, Gr. Corolla white or rose-tinged, ½-¾ in. broad. Cold bogs.

15. PYROLA, Tournefort.

Stamens connivent about the straight style, not declined: stigma peltate.......... 1, 2
Stamens and style bending downward then upward: style exserted.
 Corolla greenish white: calyx-lobes short............................ 3, 4, 6
 Corolla rose-purple or purplish: scaly bracts large........................ 5
 Leaves sometimes veined or splotched with white..... .... 5, 6
 Leaves wanting: scapes reddish: petals obovate, white..................... 7
1. P. minor, L. Leaves orbicular, an inch long or less: style short.
2. P. secunda, L. Leaves ovate, 1 to 2 inches long: petals oblong: style long.
3. P. chlorantha, Swartz. Leaves orbicular, 5 to 8 lines long: sepals obtuse.
4. P. elliptica, Nutt. Leaves 1½ to 2½ inches long, longer than the petioles.
5. P. rotundifolia, L. Leaves orbicular or nearly so, shining above. Only the
 var. bracteata, Gr., found on this coast, which often has large white-banded leaves.
6. P. picta, Smith. Leaves broadly ovate to narrow or spatulate, coriaceous.
7. P. aphylla, Smith. Scapes a span to a foot high: bracts subulate.

16. ALLOTROPA, Torrey & Gray.

1. A. virgata, T. & G. Thick and densely bracteate at base, ending in a long spike.

17. PTEROSPORA, Nuttall.

1. **P. andromedea,** Nutt. Pedicels slender, soon recurved: corolla globose, white.

18. SARCODES, Torrey.

1. **S. sanguinea,** Torr. A span to a foot high: flowers erect on thick pedicels.

19. MONOTROPA, Linnæus.

1. **M. uniflora,** L. Smooth: mostly white, rarely flesh-color: single flower nodding
2. **M. Hypopitys,** L. Tawny or flesh-color: petals 4, except in terminal flower.
3. **M. fimbriata,** Gr. Bracts and spatulate sepals lacerate-fringed: petals mostly 3.

20. PLEURICOSPORA, Gray.

1. **P. fimbriolata,** Gr. Brownish, stout: anthers opening lengthwise: ovary 1-celled.

21. NEWBERRYA, Torrey.

1. **N. congesta,** Torr. Flowers capitate: corolla-tube longer than the lobes.
2. **N. spicata,** Gr. Flowers spicate: corolla-tube broader, as long as the lobes.

LENNOACEÆ.

1. PHOLISMA, Nuttall.

1. **P. arenarium,** Nutt. Brownish or reddish stems in clumps: spike 1 or 2 inches long: purplish: corolla exceeding the linear bracts and sepals. Monterey, S.

PLUMBAGINACEÆ.

Petaloid calyx scarious, plicate: petals long clawed: styles filiform.
Leaves oblong or spatulate: scapes branching paniculately: spikes 1-sided............ 1
Leaves grass-like: simple scapes bearing a globose head of purplish flowers..:........ 2

2. STATICE, Tournefort.

1. **S. Limonium,** L. var. Californica, Gr. Lavender flowers in compound spikes.

3. ARMERIA, Willdenow.

1. **A. vulgaris,** Willd. Short-pediceled flowers surrounded by scarious bracts.

PRIMULACEÆ.

Leaves all radical: nodding flowers on a naked scape in a bracteate umbel....... .... **1**
Leaves radical or crowded on tufted stems, cuneate-spatulate, 5-7-toothed at apex.... **2**
Leaves in a whorl at top of stem, bracts below: corolla rotate, rose to white.......... **3**
Leaves all or mostly opposite: flowers axillary.
    Flowers small, yellowish, in close clusters; corolla rotate..................... **4**
    apetalous, solitary, purplish or white......................... **5**
    solitary: corolla rotate on slender pedicel...................... **6**
Leaves all or mostly alternate: flowers solitary, minute............................ **7**
    flowers in paniculate racemes, very small............ **8**

1. DODECATHEON, Linnæus.

* *Short filaments united to form with the closely connivent anthers a dark colored beak surmounting the short corolla tube.*
Capsule obtuse, splitting at or from the apex into valves.
    Leaves from narrowly to broadly spatulate: capsule oblong or longer.......... **1**
    Leaves obovate or oval, short, base cuneate: capsule globular................. **2**
Capsule cylindraceous; apex not splitting, but coming off as a lid.................... **3**
* * *Short distinct filaments included in the corolla throat, only the anthers exserted: leaves oval or ovate to oblong, not tapering at base.....................................* **4**
1   **D. Jeffreyi**, Moore. Often very large: capsule exceeding calyx.
2   **D. ellipticum**, Nutt. Leaves ½ to 2 inches long: calyx minutely glandular.
3   **D. Hendersoni**, Gr. Like the last except the thin-walled exserted capsule.
4   **D. frigidum**, C. & S. var. dentatum. Leaves commonly repand or dentate.

2. PRIMULA, Linnæus.

1. **P. suffrutescens**, Gr. Scape 2 to 4 inches long: umbel of several red-purple flowers.

3. TRIENTALIS, Linnæus.

1. **T. Europæa**, L. Flowers on slender pedicels among the leaves. Our plants are:
Var. latifolia, Torr., with leaves mostly acute, ½ to 4 inches long, and
Var. arctica, Ledeb., with obtuse or retuse leaves an inch long or less.

4. LYSIMACHIA, Tournefort.

1. **L. thyrsiflora**, L. Leaves lanceolate: small teeth between corolla lobes.

5. GLAUX, Tournefort.

1. **G. maritima**, L. Succulent, pale green, 3 or 4 inches high, leafy.

6. ANAGALLIS, Tournefort.

L. A. arvensis, L. Square stems: leaves ovate: corolla often salmon-purple.

7. CENTUNCULUS, Dillenius.

1, C. minimus, L. Slender: corolla lobes acute, shorter than calyx.

8. SAMOLUS, Tournefort.

L. S. Valerandi, L., var. Americanus, Gr. Corolla white, a line long or less.

STYRACACEÆ.

1. STYRAX, Tournefort.

L. S. Californica, Torr. Shrub: spatulate corolla lobes, 8 or 9 lines long, white.

OLEACEÆ.

1. FRAXINUS, Tournefort.

1. F. dipetala, H. & A. Leaflets serrate: petals 2, white, 2 lines long.
2. F. Oregana, Nutt. Leaflets mostly entire: flowers diœcious, apetalous.

APOCYNACEÆ.

Flowers in terminal cymes: corolla campanulate, white or pinkish..................... 1
Flowers on scape-like peduncles: corolla short: funnelform, rose-purple.............. 2

1. APOCYNUM, Tournefort.

1. A. androsæmifolium, L. Spreading: leaves ovate: corolla 3 or 4 lines long.
2. A. cannabinum, L. More strict: leaves narrower, nearly sessile: corolla smaller.

2. CYCLADENIA, Bentham.

1, C. humilis, Benth. Smooth, low: corolla 9 lines long, throat hairy: style long.
Var. tomentosa, Gr. Densely hairy: leaves 2 or 3 pairs, 1 to 3 inches long.

ASCLEPIADACEÆ.

Stem twining: anthers with scale-like appendages: corolla rotate.......................... 1
Stem erect: anthers with hooded or cup-like appendages: petals reflexed.

Hoods with horn-like process within.. 2
Hoods cleft at the back (outside), hornless..................................... 3
Hoods cleft on the inside, hornless... 4

1. PHILBERTIA, HBK.

1. **P. linearis, var. heterophylla,** Gr. Corolla 6 lines broad, dull-colored. S. Cal.

2. ASCLEPIAS, Linnæus.

Corolla-lobes 4 or 5 lines long: hoods 5 or 6 lines long, back prolonged............... 1
Corolla-lobes whitish, 3 lines long: hoods truncate; horns little exserted............. 2
Corolla-lobes greenish, 3 or 4 lines long: hoods appendaged on sides.................. 3
Corolla-lobes whitish, ovate, 3 lines long..... 4
Corolla-lobes greenish or purplish, 3 lines long: horns triangular, obtuse............ 5
Corolla-lobes greenish or purplish, 2 lines long: horns slender, exserted.............. 6
1. **A. speciosa,** Torr. Stout, 2 to 5 ft. high: follicles with soft spines.
2. **A. Fremonti,** Torr. A foot high or less: short-woolly: leaves obtuse.
3. **A. erosa,** Torr. Leaves ovate or narrower, acuminate, margins scarious.
4. **A. eriocarpa,** Benth. Densely woolly: leaves often in 3's, 4 to 8 inches long.
5. **A. vestita,** H. & A. Dense white wool deciduous in age: leaves very acute, long.
6. **A. Mexicana,** Cav. Smooth; slender leaves in whorls, 3 to 6 inches long.

3. SCHIZNOTUS, Gray.

1. **S. purpurascens,** Gr. Decumbent or prostrate: leaves cordate: corolla reddish.

4. GOMPHOCARPUS, Robt. Brown.

1. **G. cordifolius,** Benth. Smooth: loosely flowered: corolla dark purple-red.
2. **G. tomentosus,** Gr. Woolly: stem angled: corolla greenish or purplish.

GENTIANACEÆ.

Corolla from funnelform to salverform: leaves opposite.
  Corolla yellow, 4-lobed: anthers not twisted................................... 1
  Corolla red, 3-5-lobed: anthers spirally twisted in age........................ 2
  Corolla blue or white: stigma flat, nearly sessile............................. 3
Corolla rotate, 4-parted with fringed glands: leaves opposite or whorled.......... 4
Corolla campanulate: leaves alternate or radical, 3-foliolate or reniform...... 5

1. MICROCALA, Link.

1. **M. quadrangularis,** Griseb. Slender, 2 or 3 inches high: calyx 4-angled.

2. ERYTHRÆA, Renealm.

Corolla-lobes 1½ to 2½ lines long; tube much longer: anthers oblong.............. **1, 2, 3**

Corolla-lobes 3½ to 6 lines long: tube a little longer; anthers linear................ **4, 5**

1. E. floribunda, Benth. Pedicels short or none: corolla-lobes 2 lines long or less.
2. E. Muhlenbergii, Griseb. Pedicels short or 2-bracted; corolla-lobes obtuse.
3. E. Douglasii, Gr. Pedicels slender: corolla-lobes obtuse: seeds globular.
4. E. trichantha, Griseb. Flowers often corymbose, some sessile, lobes acute.
5. E. venusta, Gr. Flowers pediceled: corolla-lobes obtuse, tube yellowish.

3. GENTIANA, Tournefort.

a. Corolla without plaited folds or appendages between the lobes.

Flowers solitary on terminal peduncle, 12 to 18 lines long........................ **1, 2**

Flowers several, smaller, 5 to 7 lines long: calyx 5-cleft........................... **3**

b. Corolla with folds between the (usually 5) lobes which are prolonged into thin teeth or accessory lobes; stigmas distinct: pod on a stipe.

Annual: anthers introrse: stem leaves ovate-cordate 2 to 4 lines long................ **4**

Perennial: anthers more or less extrorse: usually a pair of bracts or leaves under the short-peduncled or sessile flower.

Stems several from one caudex, 1-2-flowered: stem-leaves connate-sheathing.

    Stems 1-flowered, 2 to 4 inches high: radical leaves rosulate............. **5**

    Stems longer: upper pair of leaves enclosing the flower................ **6, 7**

Stems many-leaved: style manifest, corolla blue or bluish.

    Corolla-lobes broad, narrowed at base; accessory lobes entire........... **8, 9**

    Corolla-lobes not narrowed at base: accessory lobes laciniate......... **10, 11**

1. G. serrata, Gunner, var. holopetala, Gr. Calyx angular, lobes keeled.
2. G. simplex, Gr. Leaves linear-oblong, 3 to 9 lines long: calyx hardly angular.
3. G. Amarella, L. var. acuta, Engelm. Stem acute-angled: capsule sessile.
4. G. Douglasiana, Bong. Cymosely branched: radical leaves rosulate.
5. G. Newberryi, Gr. Radical leaves obovate to spatulate: corrolla 18 lines long.
6. G. setigera, Gr. Stems decumbent: 1 or 3 bristles between corolla-lobes.
7. G. calycosa, Griseb. Stems erect: accessory corolla-tubes laciniate or 2-cleft.
8. G. Menziesii, Griseb. Stems slender, a ft. long or less: leaves 1½ in. long or less
9. G. sceptrum, Griseb. Stem 2 to 4 ft. high: leaves broader, 1½ to 3 in. long.
10. G. Oregana, Engelm. Corolla over an inch long, lobes roundish.
11. G. affinis, Griseb. Corolla an inch long or less, lobes ovate, acute.

4. FRASERA. Walter.

Stout, 2 to 5 ft. high: leaves not white margined................................... **1, 2**

Gray-green, 1 to 3 ft. high: leaves with cartilaginous white margins............ **3, 4, 5**

1. F. thyrsiflora, Hook. Leaves in 2's or 3's: a gland on each corolla-lobe.

2. **F. speciosa**, Dougl. Leaves in 4's and 6's: 2 glands on each corolla-lobe,
3. **F. Parryi**, Torr. Leaves in 2's or 3's: corolla white, glands lunate-obcordate.
4. **F. nitida**, Benth. Slender: light blue corolla often greenish spotted.
5. **F. albicaulis**, Dougl. Similar but minutely puberulent: glands linear-oblong.

5. MENYANTHES, Tournefort.

1. **M. trifoliata**, L. Leaves 3-foliolate: flowers racemose: corolla bearded.
2. **M. Christa-galli**, Menz. Leaves reniform: flowers cymose, crested.

POLEMONIACEÆ.

Leaves entire, opposite: corolla salverform, rose-purple to white: stamens inserted at unequal heights: perennials.. 1
Leaves various; rarely all opposite and entire, then the stamens are inserted at equal heights: corolla from salverform and funnelform to almost rotate 2
Leaves simply pinnate, alternate; leaflets entire, apex sharp: corolla rotate to funnelform: stamens declined, hairy at base.. 3

1. PHLOX. Linnæus.

Matted cushion-like, evergreen: leaves narrow, crowded, 3 to 6 lines long.
  Woolly, in mats 2 to 4 inches high: leaves imbricated, recurved............... 1
  Not woolly: leaves rigid, hispid-ciliate, sometimes recurved................. 2
  Not woolly, less densely tufted: leaves narrower, less rigid................. 3
Loosely tufted: leaves linear to ovate, mostly exceeding an inch long.
  Leaves very narrowly linear, style long, slender........................ 4, 5
  Leaves linear to ovate: corolla usually 6 to 10 lines broad............... 6, 7
1. **P. canescens**, T. & G. Corolla white, 6 to 9 lines long, tube exserted.
2. **P. cæspitosa**, Nutt. Corolla tube a little exceeding the calyx lobes.
3. **P. Douglasii**, Hook. Leaves with margins naked or ciliate at base.
4. **P. linearifolia**, Gr. Much branched: leaves 1 or 2 inches long a line wide.
5. **P. longifolia**, Nutt. Similar but lower and cells mostly 1-ovuled.
6. **P. adsurgens**, Torr. Smooth leaves ovate or narrower: corolla-tube long.
7. **P. speciosa**, Pursh. Leaves lanceolate to linear: corolla tube and style short.

2. GILIA, Ruiz & Pavon.

* *Leaves opposite, at least below, palmately parted into linear or filiform divisions (entire in 8 and rarely in 10).*
Diffusely branching to nearly simple stems: corolla nearly rotate to salverform.
  Flowers scattered on filiform pedicels................................... 1 to 8
  Flowers sessile, a few together or solitary............................. 9, 10
10

Simple or sparingly branched: flowers sessile in dense leafy-bracted heads: corolla salverform

     Corolla-tube little or not at all exserted beyond the leafy bracts...... **11, 15, 16**

     Corolla tube much exserted.................................... . **12, 13, 14**

     * * *Leaves alternate, lobed or parted; rarely a few entire or opposite.*

† *Leaves palmately parted into rigid pungent divisions: stems woody: flowers large, sessile:
         corolla salverform: stamens included.................* **17, 18**

       †† *Leaves pedately 5-7-parted: soft-hairy perennials.*

     Flowers white in dense heads: some leaves 3-parted or entire............ .. **34**

     Flowers violet or purplish, solitary, subsessile in forks or axils.............. **35**

††† *Leaves pinnately incised cleft or divided, rarely a few entire or opposite: bracts some-
       times nearly palmately cleft.*

a. *Flowers in dense leafy-bracted clusters or heads: lobes of the calyx, bracts and upper
       leaves mostly rigid and pungent.*

Much branched annuals: sometimes viscid: never woolly except in the heads: stigmas often
only 2.

     At least some of the leaves bipinnatifid.

         More or less viscid; odor disagreeable............................... **19, 20**

         Not viscid: leaf-segments filiform................................ **21 to 24**

     Leaves simply pinnatifid or many entire.

         Not viscid; bracts and calyx fine-woolly............................. **25**

         Viscid................................ .... **26 to 28**

Densely woolly, at least when young: corolla salverform: stamens exserted.

     Leaves rigid, not viscid: filaments exserted; anthers sagittate.......... **29 to 33**

     Leaves not rigid: petioles broad: flowers small, white, numerous............. **34**

     b. *Inflorescence bractless or nearly so: leaves not rigid or pungent.*

Stems from creeping rootstocks, 1 or 2 inches high............................. **35**

Flowers in long-pedunculate ovoid heads: leaf-lobes filiform.................... **36, 37**

Flowers clustered or solitary: leaf-lobes slender (except 41).................. **38 to 44**

Corolla pinkish, slender, twice as long as calyx............................. **45, 46**

* * * *Leaves entire (rarely 2 or 3 small lobes), alternate, or the lower opposite, sessile: corolla
     salverform to funnelform: stamen unequally inserted: more or less viscid annuals.*

Flowers on filiform peduncles: corolla pink, 5 to 10 lines long. **47**

Flowers in loose cluster or scattered: calyx-lobes slender.... .... **48**

Flowers in the forks and upper axils: calyx-lobes awn-like........................ **49**

Flowers in leafy-bracted capitate clusters or a few scattered.

     Calyx-lobes acute: corolla 5 lines long..................................... **50**

     Calyx-lobes obtuse: corolla 10 to 15 lines long.............................. **51**

§ 1. **Dactylophyllum**, Gray.

1. **G. uniflora**, Benth. Corolla white or pinkish, nearly rotate. W. Cal.

Var. pharnaceoides, Gr. Smaller: the flowers half as large, 3 to 5 lines broad.
2. G. pusilla, Benth. Corolla short funnelform, 2 or 3 lines long, throat yellowish.
Var. Californica, Gr.· Corolla larger, twice as long as calyx. Common form.
3. G. Harknessii, Curran. Corolla white, 1 or 2 lines long, tube equaling lobes.
4. G. Bolanderi, Gr. Corolla purplish, lobes exceeding the narrow tube.
5. G. ambigua. Tube, dark throat and lilac-purple limb, each 2 lines long.
6. G. Rattani, Gr. Less branched: corolla tube long exserted, slender. Cent. Cal.
7. G. aurea, Nutt. Diffuse: leaves hispidulous: very small: corolla yellow.
Var. decora, Gr. Corolla white or purplish, throat often dark. Cent. Cal. S.
8. G. dianthoides, Endl. Corolla lilac or purple, large, lobes fringed. S. Cal.
9. G. Lemmoni, Gr. Leaves minute: calyx lobes rigid: corolla yellow. S. Cal.

§ 2. Linanthus, Endl., Benth.

10. G. dichotoma, Benth. Smooth: corolla salverform, satiny-white, large.

§ 3. Leptosiphon, Endl., Benth.

11. G. densiflora, Benth. Stout: leaf-lobes stiff: corolla 8 to 10 lines broad.
12. G. androsacea, Steudel. Very variable: corolla·throat yellow or dark.
13. G. micrantha, Steud. Corolla very slender, usually yellow.
14. G· tenella, Benth. Leaves hispidulous-ciliate: corolla pink, throat yellow.
15. G· ciliata, Benth. Rigid, grayish-hispid: corolla rose color.

§ 4. Siphonella, Gray.

16. G. Nuttallii, Gr. Perennial: corolla white; throat broad, yellow.

§ 5. Leptodactylon, Bentham.

17. G. Californica, Benth. Corolla often 18 lines broad. Coast.
18. G. pungens, Benth. Viscid: corolla smaller. Sierra Nevada.

§ 6. Navarretia, Gray,

19. G. squarrosa, H. & A. Corolla blue to white: stamens included.
20. G. cotulæfolia, Steud. Less viscid: stamens exserted.
21. G. intertexta, Steud. Calyx and spiny bracts white and woolly at base.
22. G. Breweri, Gr, Less pungent: corolla yellow, 3 or 4 lines long.
23. G. leucocephala, Gr. Erect or branches procumbent, pale green.
24. G. prostrata, Gr. Similar; prostrate branches from a central head.
25. G. divaricata, Torr. Heads small; bracts nearly palmately cleft.
26. G. filicaulis, Torr. Small corolla similar, but stamens exserted.
27. G. viscidula, Gr. Stout, Diffuse: corolla violet to purple.

Var. heterodoxa, Gr.   Slender bracts broad, less rigid: corolla tube shorter.
28. G. atractyloides, Steud.   More rigid and viscid: mint scented.

§ 7. Hugelia, Gray.

29. G. densifolia, Benth.   Corolla violet-blue, tube much exserted.
30. G. virgata, Steud.   More slender: flowers fewer, blue or lavender.
Var. floribunda, Gr.   Corymbose branches ending in dense heads.
31. G. floccosa, Gr.   Corolla tube 3 or 4 lines long: anthers shorter.
32. G. filifolia, Nutt.   Corolla lobes a line long: anthers cordate-oval.
33. G. lutescens, Steud.   Corolla yellow, 3 lines long: pod 3-seeded.

§ 8. Elaphocera, Nuttall.

34. G. congesta, Hook.   Leaves pedately 5-7-parted, lobes 2 lines long.

§ 9. Eugilia, Bentham, Gray.

35. G. debilis, Watson.   Soft hairy: Flowers sessile among crowded leaves.
36. G. capitata, Dougl.   Flowers light blue: calyx scarcely hairy.
37. G. achilleæfolia, Benth.   Flowers violet to lavender: calyx-tips recurved.
39. G. multicaulis, Benth.   Corolla violet, 4 lines long: capsule ovoid.
40. G. tricolor, Benth.   Corolla-lobes violet or lilac, throat dark purple.
41. G. latifolia, Gr.   Corolla 9 or 10 lines long, purple with dark throat.
42. G. tenuiflora, Benth.   Corolla narrow, 7 to 9 lines long, rose and violet.
43. G. inconspicua, Dougl.   Corolla narrow, 3 to 5 lines long, variable.

§ 10. Ipomopsis, Bentham.

44. G. aggregata, Spreng.   Large corolla, scarlet to white, dotted; lobes acute.
Var. Bridgesii, Gr.   Lower, 6 to 18 inches high: corolla bright red. S. N. Mts.

§ 11. Courtoisia, Gray.

45. G. glutinosa, Gr.   Calyx rounded at base, deeply cleft: capsule globular.
46. G. heterophylla, Dougl.   Diffuse: calyx-base acute: clusters close.
47. G. capillaris, Kellogg.   Calyx small: corolla-lobes equaling throat.

§ 12. Collomia, Gray.

48. G. gracilis, Hook.   Leaves narrow; lowest opposite, broader.
49. G. aristella, Gr.   Corolla purplish 4 to 6 lines long: capsule 3-lobed. N. Cal. N.
50. G. linearis, Nutt.   Corolla lilac-purple to white, slender.
Var. subulata, Gr.   Low, much branched, flowers few in lower forks.
51. G. grandiflora, Dougl.   Corolla salmon color, 12 lines long.

8. POLEMONIUM.

Tufted, more or less viscid: corolla funnelform: alpine........................... 1, 2
Stems 1 to 3 ft. high: leaflets mostly an inch or more long...................... 3, 4
Slender, much branched: leaflets 2 to 4 lines long: annual......................... 5

1. P. confertum, Gr. Small leaflets 2-3-divided: flowers in heads, 6 to 12 lines long.
2. P. humile, Willd., var. pulchellum, Gr. Leaflets entire: flowers fewer.
3. P. cœruleum, L. Flowers blue, numerous, in a narrow naked panicle.
4. P. carneum, Gr. Corolla salmon or flesh color, often over an inch long.
Var. luteum, Gr. Corolla yellow, lobes (as in the species) broadly obovata. Or.
5. P. micranthum, Benth. Corolla whitish, nearly rotate, small.

HYDROPHYLLACEÆ.

§ 1. Ovary and pod globose, 1-celled, lined with a pair of expanded placentæ: corolla usually convolute in the bud. Herbs.
* *Stamens and style much exserted: calyx not enlarged in fruit: flowers in dense clusters or heads: leaves alternate: perennial...* 1
* * *Stamens shorter than the corolla: calyx enlarging in fruit: flowers scattered or in loose clusters: lower and sometimes all the leaves opposite: annual:*
Calyx with reflexed appendages between the lobes................................. 2
Calyx not appendaged: the lobes broad and obtuse: corolla white.................... 3
§ 2. Ovary 1-2-celled: calyx deeply parted: corolla imbricated in the bud.
Leaves all entire and opposite... 4
Leaves all or all but the lowest alternate simple or compound: style 2-cleft.
    Corolla deciduous, not yellow... 5
    Corolla persistant, yellow.. 6
Leaves mostly radical, long petioled, round-cordate, crenately 7-8-lobed.
Style and stigma entire: cymes bractless, racemose............................... 7
Leaves and 1-flowered peduncles all radical: corolla lobes 5 to 7................. 8
§ 3. Ovary completely or nearly 2-celled: styles distinct, the tips thickened: corolla imbricated not appendaged: leaves simple.
Woody at base or tufted: corolla narrow funnelform............................... 9
Shrubs; leaves thick, toothed: cymes terminal................................... 10

1. HYDROPHYLLUM, Tournefort.

1. H. capitatum, Dougl. Leaves 5-7-parted, lobes 2-3-cleft.
2. H. occidentale, Gr. Leaves 7-15-parted, lobes cleft, obtuse.
Var. Watsoni, Gr. Almost stemless, softer hairy.
3. H. Virginicum, L. Leaves bright green, nearly smooth, 3-5-parted.

2. NEMOPHILA, Nuttall.

Leaves all or nearly all opposite, seeds 5 or more.............................. 1, 2, 3
Leaves all or many alternate: stems weak: seeds 4 or less........................ 4, 5
1. N. **maculata**, Benth. Corolla white with 5 violet spots.
2. N. **insignis**, Dougl. Leaves 7-13-lobed: corolla bright blue.
3. N. **Menziesii**, H. & A. Corolla blue to white, dark dotted in center.
4. N. **aurita**, Lindl. Leaves 2 to 4 in. long, lobes and prickles retrorse: limb violet.
5. N. **parviflora**, Dougl. Leaves variable: white, dotted corolla 2 to 6 lines long.

3. ELLISIA, Linnæus.

1. E. **membranacea**, Benth. Leaves 3-9-divided: lobes mostly entire.
2. E. **chrysanthemifolia**, Benth. Leaves twice or thrice pinnatifid.

4. DRAPERIA, Torrey.

1. D. **systyla**, Torr. Silky viscid: leaves opposite, entire: stamens unequal.

5. PHACELIA, Jussieu.

 * *Leaves simple and entire or some of the lower ones with small entire lobes at the base.*
All simple and entire, narrow, the lower (and the branches) opposite............. 1, 2
Mostly simple and entire, ovate or oblong: spikes long........................... 20
Simple and entire or with 2 or 3 slender basal lobes, narrow...................... 25
Often simple and entire but lower ones usually with 1 to 3 pairs of basal lobes, all lanceo-
 late or ovate: veins simple, distinct. Ovules 4......................... 3, 4, 5
           Ovules 8 or more............. 28, 29, 30
 * * *Leaves simple and more or less notched or lobed, or lower ones with small basal lobes,*
          ovate or cordate.
Hispid with spreading stinging hairs, annual.................................... 6, 7
Hispid, viscid: leaves often pinnatifidly lobed.................................. 21
Viscid: flowers large in loose racemes, blue, violet or white.
 Very viscid: style 2-parted.............. .... 14, 15
 Less viscid: style 2-cleft; corolla blue or violet...................... 16, 17
 Leaves doubly toothed or some pinnately parted........................... 18
 Leaves small, shorter than the petioles 19
Lower leaves with small basal divisions.
 Leaves and flowers large, viscid 22
 Leaves silky: somewhat hispid and glandular.............................. 23
   * * * *Leaves 1-3-pinnately divided and incised.*
Calyx not hispid, 2 lines long in fruit: seed mostly solitary.................... 8
Calyx hispid or ciliate: style 2-parted.........................9, 10, 11, 12, 13

Style cleft to near the middle, leaves simply pinnate.

Tall perennial, soft pubescent; leaves large........... .... **24**

Leaves with 7 to 15 entire or few-toothed obtuse lobes....................... **26**

Leaves mostly at base: flowers on pedicels 6 to 12 lines long............... **27**

Style cleft at apex: corolla nearly tubular, 5 to 7 lines long....................... **31**

§ 1. Euphacelia, *Gr. Ovules 4.*

1. **P. namatoides**, Gr. A span high: corolla blue, 1 or 2 lines long.
2. **P. Pringlei**, Gr. Taller: corolla more broadly campanulate. N. Cal.
3. **P. circinata**, Jacq. f. Hispid: grayish leaves strigose: spikes dense.

Var. **calycosa**, Gr. Calyx-lobes broader, veiny: stamens as much exserted.

4. **P. Breweri**, Gr. Similar but annual, smaller: hairless filaments not exserted.
5. **P. humilis**, T. & G. Diffuse: a span high: corolla deep blue, 2 or 3 lines long.
6. **P. malvæfolia**, Cham. Corolla white, 3 or 4 lines broad: stamens exserted.
7. **P. Rattani**, Gr. More slender: corolla 2 lines long: stamens included.
8. **P. platyloba**, Gr. Corolla nearly rotate, bluish, little exceeding calyx.
9. **P. distans**, Benth. Corolla dull-white to violet: stamens scarcely exserted.
10. **P. tanacetifolia**, Benth. Similar but stamens much exserted: capsule oval.
11. **P. hispida**, Gr. White-hispid: sepals very slender, much exceeding globose capsule.
12. **P. ramosissima**, Dougl. Perennial: stems weak: leaves rather coarsely lobed.
13. **P. ciliata**, Benth. Calyx much enlarged in fruit, lobes ovate, ciliate, veiny.

§ 2. Gymnobathus, *Gr. Ovules and seeds numerous: no appendages to rotate campanulate corolla.*

14. **P. viscida**. Torr. Corolla deep blue with lighter center, 6 to 12 lines broad.

Var. **albiflora**, Gr. Flowers white. With next species. Santa Barbara, S.

15. **P. grandiflora**, Gr. Similar: light blue to white corolla much larger.

§ 3. Whitlavia ,*Gr. Ovules 8 to many: flowers showy.*

16. **P. Whitlavia**, Gr. Corolla-tube cylindrical, spreading lobes much shorter.
17. **P. campanularia**, Gr. Corolla campanulate, 8 to 10 lines long. San Diego.
18. **P. Parryi**, Torr. Corolla cleft below the middle, violet, often 5 spots in throat.
19. **P. longipes**, Torr. Slender: corolla 5 or 6 lines long, white. Los Angeles, S.

§ 4. Eutoca, *Gr. Ovules 10 to many: capsule ovoid or oblong.*

20. **P. grisea**, Gr. Corolla whitish: filaments retrorsely hairy, exserted.
21. **P. loasæfolia**, Torr. Corolla 3 lines long: naked filaments much exserted.
22. **P. Bolanderi**, Gr. Corolla nearly rotate, 10 or 12 lines broad, violet to white.
23. **P. hydrophylloides**, Torr. Corolla 3 or 4 lines broad: naked filaments much exserted.

24. **P. procera,** Gr. Leaf-lobes acute: filaments much exserted.
25. **P. Menziesii,** Torr. Corolla violet or white, 6 to 10 lines broad.
26. **P. brachyloba,** Gr. Corolla small, whitish: stamens not exserted.
27. **P. Douglasii,** Torr. Diffuse: corolla campanulate, 5 to 10 lines broad.
28. **P. Davidsoni,** Gr. Hoary: leaves strigose: pedicels equaling calyx.
29. **P. circinatiformis,** Gr. Spikes dense: stamens included: seeds 6 or more.
30. **P. divaricata,** Gr. Corolla broadly campanulate, blue, 7 to 10 lines broad.
   § 5. **Microgenetes,** *Gr. Style cleft only at apex: stamens unequal, included.*
31. **P. bicolor,** Torr. Diffuse: racemes loose: corolla-tube yellowish.

6. EMMENANTHE, Bentham.

1. **E. parviflora,** Gr. Very viscid: corolla not exceeding calyx.
2. **E. penduliflora,** Benth. Less viscid: corolla exceeding calyx.

7. ROMANZOFFIA, Chamisso.

1. **R. Unalaskensis,** Cham. Calyx-lobes little shorter than the corolla.
2. **R. Sitchensis,** Bong. Pedicels, funnelform corolla and style longer.

8. HESPEROCHIRON, Watson.

1. **H. Californicus,** Wat. Corolla-lobes shorter than the tube.
2. **H. pumilus,** Porter. Corolla nearly rotate, tube bearded within.

9. NAMA, Linnæus.

1. **N. Lobbi,** Gr. Silky-woolly: leaves entire: flowers nearly sessile.
2. **N. Rothrockii,** Gr. Leaves almost pinnatifid: flowers in terminal heads.
3. **N. Parryi,** Gr. Cymes scorpioid: leaves linear, undulate, villous.

10. ERIODICTYON, Bentham.

1. **E. tomentosum,** Benth. Whitened or rusty with dense pubescence. S. Cal.
2. **E. glutinosum,** Benth. Sticky, resinous coated: corolla 6 lines long. Cal.

BORRAGINACEÆ.

§ 1. Ovary merely 4-lobed: stigma broad, sessile: glabrous: succulent................ 1
§ 2. Ovary 4-parted into seed-like nutlets; style conspicuous; stigma small.
    * *Nutlets fixed by the base to a flat receptacle, smooth and shining.*
    Flowers leafy-bracted: corolla imbricated, yellow: soft-hairy.................. 2
    Flowers bractless: corolla convolute, blue or white........................... 3

* * *Nutlets fixed to a prominent base (gynobase) by some part of the inner angle or face:*
corolla imbricated.
Nutlets not armed with prickles, not appendaged.
    Corolla blue or whitish: smooth glaucous perennials........................... 4
    Corolla yellow: hispid annuals.. 5
    Corolla white, mostly yellow-crested in the throat: hirsute or hispid.
        Nutlets erect and straight: calyx in fruit not rotate..... 6
        ·Nutlets oblique or incurved on a rounded base........................ 7
    Corolla blue, rotate: a dwarf alpine tufted perennial........................ 8
Nutlets armed with hooked or barbed prickles, or flat and wing-margined.
    Corolla blue, purple or white; throat with a ring of 2-lobed crests.
        Racemes bracteate at base: nutlets erect, prickles barbed............... 9
        Racemes on naked peduncles: nutlets globose........................ 10
    Corolla minute, white: flowers scattered along leafy branches.
        Nutlets flattened, forming an x-shaped or star-like bur................ 11

1. HELIOTROPIUM, Tournefort.

1. H. Curassavicum, L. Nearly or quite prostrate: corolla bluish or white.

2. LITHOSPERMUM, Tournefort.

1. L. Californicum, Gr. Corolla 9 or 10 lines long: throat exceeding lobes.
2. L. pilosum, Nutt. Corolla greenish yellow, silky, 5 or 6 lines long.

3. MYOSOTIS, Linnæus.

1. M. verna, Nutt. Hispid calyx unequal: corolla white, small. Oregon.
2. M. sylvatica, Hoffm. var. alpestris, Koch. Corolla blue, 3 or 4 lines broad.

4. MERTENSIA, Roth.

1. M. maritima, Don. Corolla 3 or 4 lines long, tube shorter than calyx.
2. M. Siberica, Don. Corolla-tube much exserted: calyx lobes obtuse.

5. AMSINCKIA, Lehmann.

Nutlets sharply 3-angled, straight, smooth, shining....... 1
Nutlets broad; the back nearly flat, wavy-wrinkled cross-wise.............. ..:...... 2
Nutlets incurved, convex and ridged on the back, rough................... 3, 4, 5
1. A. vernicosa, H. & A. Sparingly hispid: corolla-tube a little exserted.
Var. grandiflora, Gr. Very bristly-hispid: corolla-tube longer, limb broader.
2. A. tessellata, Gr. Coarsely hispid: leaves mostly obtuse: calyx rusty.
3. A. intermedia, F. & M. Calyx whitish or tawny hispid: corolla 2 or 3 lines.
4. A. spectabilis. F. & M. Corolla bright orange much exserted.

BORRAGINACEÆ.

5. A. lycopsoides, Lehm. Stiff bristles with pimple-like base: leaf margins often undulate: often branching: very variable.

6. KRYNITZKIA, Fischer & Meyer.

§ 1. Nutlets ovoid, smooth, shining, a ridge down the back, a groove down the inner side, attached to the gynobase one quarter the length.......................... 1

§ 2. Nutlets ovoid, somewhat rugose, a ridge down the inner side, fixed by the base of the inner angle. Entire plant light green.

* *Mostly diffuse: lower leaves often opposite: corolla 1 or 2 lines broad.............. 2, 3

* * *Flowers numerous: limb of corolla nearly rotate, 3 to 5 lines broad: yellow crests in the throat conspicuous: lower leaves mostly opposite (except in No. 6)............ 4, 5, 6

§ 3. Nutlets never rugose; inner angle furrowed from less than half to all the way; back convex; side angles mostly obtuse, never margined: calyx in fruit erect or closed: corolla small, throat naked or the crests not exserted: numerous flowers sessile in scorpioid spikes.

* *Fruiting calyx often falling with the enclosed nutlets, these smooth, shining, acute: sepals narrow, hispid, slender.

Nutlets solitary, rarely 2, acuminate, fixed below the middle................. 7, 8, 9

Nutlets usually all maturing scarcely a line long............................. 10, 11

Nutlets unequal, one much larger than the others................................ 12

Nutlets 3-angled-ovoid, papillose, sharply muricate or scabrous, attached nearly or quite up to the apex: usually erect and hispid; spikes bractless: calyx pungent-bristly.

Calyx very villous-hispid, in fruit 3-5 lines long, mid-rib strong.................. 13

Calyx 3 lines long or less; bristles pungent, whitish or yellowish.

    In fruit double the length of the nutlets not connivent................. 14, 15

    In fruit 1 or 2 lines long, more or less connivent over the angular nutlets..16, 17, 18

* * *Fruiting calyx deciduous above a persistent basal cup: nutlets ovate-deltoid, 3-angled, usually very smooth, groove forked.

Much branched, with flowers almost from base, hispid............................. 19

1. K. lithocarya, Greene. Corolla not surpassing the rusty calyx: spike simple.
2. K. Californica, Gr. Leaves small, narrow: flowering from near the base.
Var. subglochidiata, Gr. Succulent: nutlets minute-bristly with barbed hairs.
3. K. trachycarpa, Gr. More lower leaves opposite: nutlets broader, granulate.
4. K. Chorisiana, Gr. Some pedicels 2 to 12 lines long: leaves large.
5. K. Scouleri, Gr. Slender: spikes often branching mostly bractless.
5. K. mollis, Gr. Perennial stems creeping, soft-hairy. Wet borders of ponds.
7. K. sparsiflora, Greene. Sepals with stiff hooked bristles: nutlet flattened.
8. K. oxycarya, Gr. Strigulose: leaves linear: calyx in fruit deflexed-bristly at base
9. K. microstachys, Greene. Smaller, hispidulous: calyx bristles not deflexed.
10. K. leiocarpa, F. & M. Nutlets attached for nearly the whole length.
11. K. Torreyana, Gr. Nutlets attached half way up, groove forked.

Var. calycosa, Gr.   Flowers crowded, somewhat capitate: calyx longer.
12. K. dumetorum, Greene. Almost climbing: papillose-hispid: 2 sepals united.
13. K. barbigera, Gr. Nutlets gray, very rough, rarely all fertile.
14. K. intermedia, Gr. Nutlets thickly muricate, groove with open basal scar.
15. K. ambigua, Gr. Nutlets minutely muricate, groove widely forked.
16. K. muriculata, Gr. Stout: spikes 2-3-radiate; nutlets triangular-ovate.
17. K. Jonesii, Gr. Slender: spikes more numerous, paniculate: calyx smaller.
18. K. micromeres, Gr. Hispid, diffuse: spikes filiform: flowers minute.
19. K. micrantha, Gr. var. lepida, Gr. Roots red: hispid: corolla 2½ lines long.

7. PLAGIOBOTHRYS, Fischer & Meyer.

* *Nutlets not on stipe-like attachments: calyx more or less villous with yellowish or rusty hairs, sometimes deciduous above the base (circumscissile).*
Sepals nearly distinct; in fruit 3 lines long, lax: nutlets broadly ovate............... 1
Calyx deeply 5-cleft: giving a violet stain to paper........................... 2, 3, 4
Calyx cleft nearly to the base, 2-3 lines long in fruit not connivent.................. 5
Calyx cleft half way, silky, in fruit connivent, soon circumscissile................... 6
* * *Nutlets on stipe-like attachments: hispidulous*.................................. 7
1. P. rufescens, F. & M. Stems slender from rosulate tuft of radical leaves.
2. P. tenellus, Gr. Radical leaves rosulate: nutlets 4-lobed or cross-like, shining.
3. P. Shastensis, Greene. Similar, with larger flowers and nutlets. Mt. Shasta.
4. P. Torreyi, Gr. Diffusely procumbent, hispidulous: leaves oblong.
5. P. canescens, Benth. Villous: spikes, as in the last, often leafy below.
6. P. nothofulvus, Gr. Rosulate leaves thin: corolla 2 or 3 lines broad.
7. P. Cooperi, Gr. Diffuse: corolla 2 or 3 lines broad, throat closed.

8. OMPHALODES, Tournefort.

1. O. Howardi, Gr. Silky, silvery: flowers few: corolla 4-5 lines broad. Or.

9. ECHINOSPERMUM, Lehmann.

Prickles of the fruit barbed at apex only: calyx in fruit reflexed............... 1, 2, 3
Prickles barbed to the base: crests of small white corolla small...................... 4
1. E. Californicum, Gr. Corolla short-funnelform, blue, 2-6 lines broad.
2. E. floribundum, Lehm. Corolla rotate, blue or often white, 2-3 lines broad.
3. E. diffusum, Lehm. Similar corolla 4-9 lines broad: back of nutlet naked.
4. E. Greenei, Gr. Diffuse: nutlets triangular-ovoid: prickles terete. N. Cal.

10. CYNOGLOSSUM, Lehmann.

1. C. occidentale, Gr. Hispidulous: upper leaves sessile; lower, spatulate.
2. C. grande, Dougl. Soft-villous becoming glabrate: leaves all petioled.
Var. læve, Gr. Smooth: corolla smaller, lobes shorter than tube.

156

CONVOLVULACEÆ.

11. PECTOCARYA, De Candolle.

Nutlets forming an x-shaped bur, the wings undulate or laciniate 1, 5
Nutlets forming a flat +-shaped bur, the thin margins entire 3, 4
1. P. linearis, DC. Wings of nutlets toothed, the teeth bristle-tipped.
2. P. penicillata, A. DC. More diffuse: nutlets fiddle-shaped; apex bristly.
3. P. setosa, Gr. Hispid, stouter: calyx-lobes with 3 or 4 very large bristles.
4. P. pusilla, Gr. Strigulose: nutlets angular, flat, wingless, with a midnerve.

CONVOLVULACEÆ.

Twining or trailing: corolla funnelform, large, limb entire: stigmas 2 1
Not twining: corolla 2-3 lines long, 5-cleft, white: styles 2 2
Corolla ⅓ in. long, 5-cleft, purplish: stigmas 2 .... Sp. 7 in No. 1
Twining leafless thread-like orange or yellowish stems: parasitic 3

1. CONVOLVULUS, Linnæus.

Solitary flower with a pair of broad bracts enclosing the calyx.
    Stems very short and erect or prostrate, trailing (See var. No. 5) 1, 3, 4
    Stems twining freely: bracts cordate-ovate or sagittate (See 5) 5
Flowers often 2-3 together with small bracts; stems often woody 6
Flowers with a pair of subulate bracts at base of pedicel: stamens slender 6
Flowers 3 lines long, deeply 5-cleft: not twining 7
1. C. Soldanella, L. Glabrous, fleshy: leaves reniform: flowers pinkish.
2. C. sepium, L., var. Americanus, Gr. Leaves acute: corolla rose,
3. C. Californicus, Choisy. Short, erect, or at length prostrate; pubescent.
4. C. villosus, Gr. Densely white-velvety: leaves an inch long or less.
Var. fulcratus, Gr. Bracts similar to the leaves (hastate): corolla yellowish.
5. C. occidentalis, Gr. Bracts variable: stems often very long: corolla white.
Var. tenuissimus, Gr. Only a ft. or a yd. high: leaves slender-hastate.
6. C. arvensis, L. Creeping in moist places: corolla an inch long or less.
7. C. pentapetaloides, L. A diffusely branched slender annual with spatulate or
lanceolate mostly entire leaves. This with No. 6 naturalized from Eu.

2. CRESSA, Linnæus.

1. C. Cretica, L. Gray silky leaves numerous, small, entire: flowers axillary.

3. CUSCUTA, Tournefort.

Ovary and capsule depressed glabose: stamens low, very slender................ 1, 2
Ovary and capsule pointed; corolla withering-persistent: scales fringed.

Stems slender, low, growing on salt-marsh plants.. 8
Stems coarser: corolla much exceeding the calyx, 2½ to 4 lines long.................. 4
Stems coarse: corolla lobes acute, usually inflexed............................... 5, 6
1. C. arvensis, Beyrich. Flowers scarcely a line long in dense clusters.
2. C. Californica, Choisy. Calyx lobes acute: corolla lobes slender.
3. C. salina, Engelm. Delicate white flowers 1½-2½ lines long.
4. C. subinclusa, D. & H. Flower clusters globose, 6 to 12 lines thick.
5. C. decora, Choisy. Flowers fleshy and papillose: clusters close.
6. C. racemosa, Martius, var. Chiliana, Engelm. Corolla thin. From Chili.

SOLANACEÆ.

Corolla rotate: anthers connivent around the style: fruit a berry...................... 1
Corolla rotate-campanulate: anthers not connivent: berry enclosed in the loose inflated
    calyx. May appear in cultivated ground. *Physalis.*
Corolla tubular-funnelform, ½ in. long or less: spiny shrubs........................... 2
Corolla funnelform, large: fruit large, covered with spines........................... 3
Corolla funnelform or tubular: capsule small, smooth 4
Corolla funnelform, 3-5 lines long: limb purple: stamens unequal..................... 5

1. SOLANUM, Tournefort.

Corolla deeply 5-cleft, white or bluish: berries black or red...................... 1, 2
Corolla only 5-angled or slightly lobed, violet or blue........................... 3, 4
1. S. nigrum, L. Herbaceous: leaves mostly ovate, sinuate-toothed, acute.
2. S. Douglasii, Dunal. Woody-stemmed: flowers 5-8 lines broad, often bluish.
3. S. Xanti, Gr. Woody at base: hairs jointed: leaves mostly obtuse at base.
4. S. umbelliferum, Esch. More woody: hairs branched: leaf-base mostly narrow.

2. LYCIUM, Linnæus.

1. L. Californicum, Nutt. Leaves fleshy, 1-3 lines long: flower parts in 4's.
2. L. Andersoni, Gr. Leaves larger: flowers larger, 5-6 lines long. San Diego.

3. DATURA. Linnæus.

Calyx acutely 5-angled: fruit erect, ovoid..................................... 1, 2, 3
                        Fruit nodding, globose............................... 4
Calyx scarcely angled: corolla 5 or 6 in. long: fruit nodding globose................. 5
1. D. Stramonium, L. Smooth: corolla white: lower prickles of capsule shorter.
2. D. Tatula, L. Similar: stems usually purple: corolla pale violet.
3. D. quercifolia, HBK. Leaves sinuate pinnatifid: prickles unequal, flat.

4. **D. discolor, Bernh.** Corolla purplish: capsule and stout prickles pubescent.
5. **D. meteloides, DC.** Leaves entire or repand, one-sided: capsule large. S. Cal.

4. NICOTIANA, Tournefort.

Very viscid, ill-scented herbs: flowers soon closing in sunshine.
    Corolla salverform, the limb 4 or 6 lines broad.............................. 1,
    Corolla tubular-funnelform, white; stamens unequally inserted.............. 3,
Very smooth glaucous shrub: corolla tubular, greenish yellow......................
1. **N. Clevelandi, Gr.** Corolla greenish white, violet tinged, an inch long. S. Cal.
2. **N. attenuata, Torr.** Calyx shorter: corolla longer, limb white.
3. **N. Bigelovii, Watson.** Corolla 1-2 inches long, nearly as broad.
Var. **Wallacei, Gr.** Corolla smaller: leaves often nearly clasping. S. Cal.
4. **N. quadrivalis, Pursh.** Corolla broader than long: capsule globular. Or. E.
Var. **multivalis, Gr.** Corolla often 2 in. broad, 5-8-lobed: capsule large. Or.
5. **N. glauca, Graham.** Leaves long-petioled, subcordate. Naturalized. S. Cal.

5. PETUNIA, Juss.

1. **P. parviflora, Juss.** Spreading or prostrate on the sea shore.   .

SCROPHULARIACEÆ.

A. LEAVES all or all but the lower ones alternate, rarely all radical.
* *Leaves simple and entire (except in No. 1 and 16): corolla sometimes nearly regular, upp*
        lip not beak-like.
Corolla 5-lobed, rotate: stamens 5; filaments woolly................................
Corolla bilabiate, throat nearly closed: stamens 4.
    Base of corolla prolonged into a slender spur on lower side................. ....
    Base of corolla swollen or saccate on lower side.... ....
Corolla large, open bilabiate with dentate lobes: stamens 2.........,.................
Corolla nearly regular: stamens 4: leaves narrow, mostly radical................... 1
Corolla large, campanulate-bilabiate, 4-lobed: stamens 4......................... 1
Corolla and calyx 4-lobed: flowers small: stamens 2.
    Leaves cordate-orbicular, all radical, scapes slender........................ 1
    Leaves on the stem: corolla rotate, 4-lobed.................................. 1
* * *Leaves or at least the bracts incisely lobed or pinnate: corolla tubular-bilabiate, closed*
        upper lip beak-like or compressed on the sides.
Leaves or lobes not serrate: anther cells unequal or only one.
    Upper corolla-lip much surpassing the 3-toothed obscure lower lip........... .. 1
    Upper lip erect, much smaller than the 3-saccate, 3-toothed lower lip........ 1
    Lips of club-shaped corolla nearly equal: calyx 1-3-leaved................... 2

Leaves or lobes serrate: anthers equally 2-celled.
   Upper lip or its beak exceeding the 3-toothed lower lip (except sp. 7)......... **21**
B. LEAVES all opposite or whorled (rarely alternate above in No. 9).
   * *Stamens 4 with anthers; sterile filament often rudimentary or none.*
Corolla declined: stamens and style infolded by lower lip.......................... **5**
Corolla small, lobes spreading: upper leaves 3-lobed or parted...................... **6**
Corolla erect, front lobe reflexed: scale in throat on upper side.................... **7**
Corolla etc. as in No. 9, but the seeds winged: odor rank.......................... **8**
Corolla open: sterile filament conspicuous: stigma entire........................... **9**
Corolla-throat open or closed: no ster'le filament: calyx 5-angled (except 1 sp.) **10**
Corolla blue or white, tube short, li, s spreading, the upper emarginate............ **11**
   * * *Stamens 2 with anthers: flowers small: growing in wet ground.*
Calyx 5-parted into narrow, nearly equal divisions: corolla bilabiate.
   Sterile filaments simple or none: corolla small, whitish..................... **12**
   Sterile filaments forked: corolla violet or bluish............................. **13**
Calyx 4-parted: corolla rotate, 4-lobed.. **17**

1. VERBASCUM, Linnæus.

1. **V. Thapsus, L.** Densely velvety-woolly, leaves decurrent: corolla yellow.
2. **V. virgatum, With.** Slender, green: filaments violet bearded or woolly.
3. **V. Blattaria, L.** Similar but pedicels solitary and longer than calyx.

2. LINARIA, Tournefort.

1. **L. Canadensis, Dumont.** Straight, smooth: leaves narrow: flowers blue.

3. ANTIRRHINUM, Tournefort.

* *Erect, 2 to 7 ft. high, leafy: flowers in a dense spike, light rose color: filaments broadest at the top.*
Stems many from a perennial base, simple, glabrous, light green **1**
Stem stout, branching, very viscid-pubescent, 3-5 ft. high......................... **2**
* * *Branching with filiform axillary branchlets which coil around objects: sepals unequal.*
Flowers in a more or less villous-viscid spike: bracts minute **5, 6, 7**
Flowers scattered along the stem and slender branches.
   Leaves on the main stem, ovate or subcordate........................... **8, 9**
   Leaves on main stem mostly narrow: corolla purple................... **10, 11**
* * * * *Erect, nearly simple stems: peduncles slender, twisting around objects.......... **12**
1. **A. virga, Gr.** Deflexed lower lip of corolla upward inflexed from middle.
2. **A. glandulosum, Lindley.** Corolla with yellowish palate. Monterey S.
3. **A. cornutum, Benth.** Filaments all broadest at top. Sac. Val., rare.
4. **A. leptaleum, Gr.** Leaves rarely linear: style shorter than pod. Sac. Valley S.

5.   A.   Coulterianum, Benth.   Leaves linear to oval, distinct: spike dense.
6.   A.   Orcuttianum, Gr.   More slender: spike loose: corolla smaller, 4 lines long.
7.   A.   Nevinianum, Gr.   Similar, but seeds ribbed not honeycomb-pitted.
8.   A.   subcordatum, Gr.   Leaves sessile, each subtending a flower and branchlet.
9.   A.   Nuttalianum, Benth.   Leaves petioled: pedicels often long as violet corolla.
10.   A.   vagans, Gr.   Very diffuse: broad upper sepal equaling corolla-tube.
Var. Bolanderi, Gr.   Leaves orbicular on branchlets: upper sepal broader.
11.   A.   Breweri, Gr.   Similar: slender corolla-tube exceeding upper sepal.   N. Cal.
12.   A.   strictum, Gr.   Corolla violet-purple, palate hairy.   Santa Barbara.

4. MOHAVEA, Gray.

1.   M.   viscida, Gr.   Very viscid: lower leaves opposite: corolla yellow, purple dotted.

5. COLLINSIA, Nuttall.

     * *Flowers on short pedicels or sessile in axillary whorls, 6-8 lines long.*
Corolla strongly declined, the throat as broad as long, nearly or quite at right angles with
     the short tube: gland sessile... 1, 2
Corolla less declined, throat narrower, leaves obtuse...................... 3, 4, 5
     * * *Flowers but little longer or not as long as the pedicels.*
Corolla usually more than 5 lines long, strongly declined sepals acute.......... 6, 7, 8
Corolla mostly less than 4 lines long: lips nearly equal: stems slender.......... 9 to 12
1.   C.   bicolor, Benth.   Upper corolla lip recurved, paler than the violet lower lip.
2.   C.   tinctoria, Hartweg.   Stains brown: corolla purple-striped, upper lip very short.
3.   C.   bartsiæfolia, Benth.   Leaves crenate, obtuse: calyx often white-hairy.
4.   C.   corymbosa, Herder.   Branching: flower clusters nearly capitate.   N. Cal C'st.
5.   C.   Greenei, Gr.   Corolla rich violet or lavender; upper lip short; side lobes small.
6.   C.   grandiflora, Dougl.   Flowers in whorls of 3 to 9: lower lip deep blue or violet.
Var. pusilla, Gr.   Small form: corolla only 4 or 5 lines long, deeply colored.
7.   C.   sparsiflora, F. & M.   Slender: only upper flowers in 3's, ½-⅔ in. long.
Var. divaricata.   Only 2 or 3 in. high: flowers smaller, solitary.   S. F. Bay.
8.   C.   linearis, Gr.   Paniculately branched: leaves very slender: pale corolla dark-
     dotted.   N. Cal.
9.   C.   parviflora, Dougl.   Often diffuse: corolla little exserted, 2 or 3 lines long.
10.   C.   Rattani, Gr.   Stem strict, mostly simple: corolla lips violet, 1-2 lines long.
11.   C.   Childii, Parry.   Stem similar: corolla light blue.   In forests S. Cal.
12.   C.   Torreyi, Gr.   Much branched: flowers in 3's and 6's, blue or violet.

6. TONELLA, Nuttall.

1.   T.   collinsioides, Nutt.   Diffuse: flowers on slender pedicels, a line long.

7. SCROPHULARIA, Tournefort.

l. 8. Californica, Cham. Stems square: flowers dull purple, 3-4 lines long.

8. CHELONE, Linnæus.

l. C. nemorosa, Dougl. Corolla violet-purple. In woods, Or., northward.

9. PENTSTEMON, Mitchell.

§ 1. Anther-cells soon widely separating at base, united more or less completely at top, splitting open nearly or quite the whole length.

* *Anthers densely woolly, becoming shield-shaped after shedding pollen* 1

* * *Anthers glabrous, splitting through the apex and spreading out: stems branching and shrubby, at least below: leaves leathery or parchment-like, mostly small, and short-petioled: filaments all hairy at base.*

Corolla scarlet, narrow-tubular, its upper lip erect and the lower more or less spreading: sterile filament bearded down one side.

    Leaves subcordate or ovate acutely toothed, 1 in. long or less 9

    Leaves oblong or oval ½ to 2 in. long, often canescent 3

    Leaves slender, rigid, acutely toothed, glaucous 4

Corolla yellow or yellowish, purple-tinged, not an inch long, the gaping limb longer than the tube (except No. 7): upper lip concave, lower recurved.

    Leaves lanceolate or oblong-lanceolate, denticulate 5

    Leaves similar, yellowish green, remotely serrate 6

    Leaves spatulate or oval, entire, 6 lines long or less 7

* * * *Anthers with spreading distinct cells splitting from base nearly to the top: corolla scarcely bilabiate, blue or purple.*

Leaves mostly oblong-lanceolate, glaucous 8

* * * * *Anthers splitting open from base through the united apex.*

Glaucous or pale and glabrous: leaves leathery or thick: corolla 9-12 lines long.

    Stems thick, 1 to 3 ft. high: leaves mostly ovate-lanceolate, entire 9

    Taller: leaves thinner; upper pairs acuminate, united, acutely dentate 10

    Similar but leaves thicker: corolla crimson, 9 lines long, throat narrow 11

    Leaves 1½ to 4 in. long, the upper often united: corolla cream-white, pinkish.. 12

Corolla 8 lines long or less (more in 13): thyrsus viscid in 13, 14, 16, 18.

    Corolla somewhat bilabiate lower lip and sterile filament hairy... 13, 14, 15, 16

    Corolla funnelform or tubular: sterile filament nearly or quite naked...... 17, 18

§ 2. Anthers sagittate or horse-shoe shaped, the cells opening by a continuous cleft around the apex which reaches about half way to the bases of the cells, these remaining closed and saccate, sometimes hairy but never woolly: corolla some shade of purple and blue, from rose purple to lavender (scarlet red in the last).

* *Soft-pubescent, viscid, stout: radical leaves 6-8 inches long* 19

11

162   SCROPHULARIACEÆ.

* * *Glabrous, or inflorescence puberulent or viscid: leaves toothed or pinnatifid: sterile filament
     hairy: corolla funnelform, moderately bilibiate.*
Corolla over an inch long, lobes and all the stamens sparsely hairy................. 20
Corolla an inch long or usually less 21, 22, 23
          * * * *Glabrous or puberlent: leaves all entire.*
Corolla 6 lines long, slender: sterile filaments bearded............................ 24
Corolla short bilabiate, 8 to 18 lines long: sterile filament naked.
     Calyx glandular or viscid: leaves lanceolate to spatulate................. 25, 26
     Calyx not glandular or viscid: thyrsus usually narrow.............. 27, 28, 29
1. P. **Menziesii**, Hooker. Leaves 3-12 lines long: corolla violet to pink. Cal. N.
Var. Newberryi, Gr. Corolla rose-purple or pink. Southern Sierras.
2. P. **cordifolius**, Benth. Climbing over bushes, very leafy. San Luis Obispo. S.
3. P. **corymbosus**, Benth. Erect, 1 or 2 ft. high, leafy: cyme corymbose.
4. P. **ternatus**, Torr. Branches slender: upper leaves in 3's. Kern Co. S.
5. P. **breviflorus**, Lindl. Sterile filament naked. Sierra Nevada.
6. P. **Lemmoni**, Gr. Paniculate: sterile filament yellow bearded. N. Cent. Cal.
7. P. **antirrhinoides**, Benth. Branched, leafy, paniculate: corolla pale yellow.
8. P. **glaber**, Pursh. Wide corolla 1 to 1½ in. long. Sierras eastward.
9. P. **centranthifolius**, Benth. Very glaucous: corolla tubular, bright scarlet.
10. P. **spectabilis**, Thurber. Corolla rose-purple or lilac, the limb violet.
11. P. **Clevelandi**, Gr. Corolla crimson, 9 lines long: sterile filaments bearded.
12. P. **Palmeri**, Gr. Corolla 8-9 lines broad: sterile fil. densely yellow-bearded.
13. P. **Rattani**, Gr. Leaves 3-8 in. long, denticulate: corolla pale purple. N. W. Ca.
Var. minor, Gr. Smaller: corolla 6-7 lines long. Klamath and Trinity R.
Var. Kleei, Gr. Between the foregoing in size. High peak near Santa Cruz.
14. P. **pruinosus**, Dougl. Pubescent: corolla deep blue, hairy. Or. Wash.
15. P. **ovatus**, Dougl. Pubescent: leaves ovate, serrate, green: corolla purple blue.
16. P. **confertus**, Dougl. Thyrsus in 2-5 dense whorls: corolla yellowish, small.
17. P. **deustus**, Dougl. Tufted, woody at base: corolla yellow to dull white.
18. P. **heterodoxus**, Gr. Leaves obtuse, entire. Near Donner Pass, Cal.
19. P. **glandulosus**, Lindl. Corolla lilac: sterile filaments naked. Or. Wash.
20. P. **venustus**, Dougl. Leaves closely serrate: sepals small. Or.
21. P. **diffusus**, Dougl. Often diffuse: leaves unequally serrate. Or. Wash.
22. P. **Richardsoni**, Dougl. Leaves incised or laciniate-pinnatifid. Or.
23. P. **triphyllus**, Dougl. Leaves lanceolate or linear, rigid, often laciniate.
24. P. **gracilentus**, Gr. Peduncles and calyx viscid. Mts. N. Cal., Or.
25. P. **lætus**, Gr. Ashy-pubescent: corolla an inch long blue. Mts. Cal.
26. P. **Roezli**, Regel. Smaller: corolla smaller, paler. Sierra Nev. to Or.
27. P. **azureus**, Benth. Glaucous; leaves ovate or narrower: corolla broad.
Var. Jaffrayanus, Gr. Low broad-leaved form in the Sierras.
Var. parvulus, Gr. Broad leaves an inch or less long: corolla 9 lines long. Alpine.

Var. angustissimus, Gr. Leaves very slender. Yosemite Valley, etc.
28. P. heterophyllus, Lindl. Similar: buds often yellowish. W. Cal.
29. P. Bridgesii, Gr. Thyrsus one-sided: corolla lips long. S. Sierras.

10. MIMULUS, Linnæus.

* *Corolla buff, salmon-color or orange, large: a viscid shrub*........................... **1**
* * *Corolla-limb rose or crimson-purple (scarlet in No. 17): sticky viscid or slimy (less so in
   8, 10, 17, 18, 38); often ill scented.*
  a. Style pubescent above; stigma unequally lobed or entire, usually peltate-funnelform:
     flowers sessile or nearly so.
Corolla-tube long, slender; lower lip very short; upper lip erect...................... **5**
Corolla scarcely exserted, 3-4 lines long: capsule much exserted...................... **6**
Corolla exceeding ½ inch; lower lip shorter; throat dark or yellow................... **8**
Corolla trumpet shape, 6-9 lines broad, crimson: calyx hardly oblique.............. **10**
Corolla similar, 6 lines long, 4-5 lines broad, deep red: calyx oblique............... **11**
Corolla nearly funnelform, 2-6 lines long, crimson: calyx-teeth spreading........... **12**
Corolla 6-9 lines long: calyx-teeth obtuse, nearly equal.............................. **13**
Corolla often an inch long: calyx-teeth very unequal, acute: very viscid............ **14**
Corolla oblique-salverform, white, crimson-veined................................ **16**
  b. Style smooth; stigma of 2 equal flat lobes which upon irritation close: flowers on long
     or short peduncles.
Corolla oblique-bilabiate, exceeding 1 inch, lobes reflexed, scarlet.................. **17**
Corolla open-bilabiate, 1½-2 inches long, lobes spreading, rose-color................ **18**
Corolla little surpassing calyx; very slimy-villous................................. **34**
Corolla 2-3 lines long; lower lip entire, upper 2 lobed............................. **38**
    * * * *Corolla-limb rose or crimson-purple: not viscid, or very slightly so.*
  a. Almost stemless: corolla-tube long and slender: style pubescent.
Flowers erect, sessile, surpassing the leaves, 1-2 inches long........ .... **2, 3, 4**
  b. Stems much longer than the flowers: style smooth; stigma equally 2-lobed, lobes flat
     and often closed... **18, 30, 38, 39**
    * * * * *Corolla yellow, often spotted: viscid or slimy.*
Corolla-throat often purple-tinged or dotted: odor strong, fetid...................... **7**
Corolla 1 inch long or longer, nearly as broad, lobes subequal...................... **15**
Corolla 3-9 lines long: peduncles scape-like: leaves rosulate-crowded **19**
Corolla ⅔-1 inch long: slimy, musky, spreading and creeping....................... **20** `
Corolla ½-⅔ inch long: fruiting calyx ½ inch long, mouth closed.................... **27**
Corolla ⅓-½ inch long: fruiting calyx on long peduncle, lower teeth shortest........ **29**
Corolla light yellow, limb often pinkish: petioles margined....................... **31**
Corolla as broad as long (½ inch): peduncles much exceeding oval leaves........... **32**
Corolla narrower, ⅓-½ inch long: peduncles little exceeding the leaves.............. **33**

Corolla-tube narrow, exserted; throat and bearded lip spotted...................... 36
Corolla 3-4 lines long, lobes nearly equal, often a pair of spots...................... 40
* * * * * * *Corolla yellow, often spotted, not viscid.*
Leaves ovate to oblong: rootstocks tuber-bearing: corolla ½ inch broad.............. 21
Leaves similar, coarsely serrate, acute: corolla orange-yellow, 1 inch broad.......... 22
Lower leaves broad, acutely and irregularly dentate or laciniate.................... 23
Lower leaves narrow, petioled, thick, shining, denticulate, small................... 24
Lower leaves clasping, the others orbicular-perfoliate, glaucous................... 25
Leaves mostly basal: stem wing-angled: upper calyx-tooth prominent.............. 26
Leaves very small, often purplish: diffuse: peduncles spreading.................... 27
Leaves narrow, laciniately lobed: corolla pale, 2-4 lines long: diffuse............... 28
Leaves narrow, entire: corolla ½ in. broad, purple dotted; lip bearded.............. 36
Leaves narrow, entire: corolla 2-3 lines long: lobes all notched..................... 37
Leaves entire, soft-hairy: diffuse: corolla with 2 brownish spots................... 40
* * * * * * *Corolla white or white and yellow, often purple-marked.*
White or yellowish, throat with 8 or 10 purple stripes............................ 9
White purple or yellowish, 3-6 lines long: calyx-teeth very short.................. 30
White, veined with crimson, oblique salverform................................. 16
Yellow with white or pinkish border, ½ in. long: viscidulous...................... 31
Upper lip white, lower yellow, purple dotted: viscid............................. 35

§ 1. **DIPLACUS**, Gr. Shrubs 3-5 ft. high: glutinous-viscid.

1. **M.** glutinosus, Wendl. Variable. Common in Central and W. Cal.

§ 2. **ŒNOE**, Gr. Corolla 1-2 in. long; tube long exserted, slender: capsule 1-sided.

2. **M.** tricolor, Lindl. Corolla limb with 5 crimson spots, palate yellow.
3. **M.** angustatus, Gr. Similar: corolla tube 3-8 times as long as the short throat.
4. **M.** Douglasii, Gr. Upper corolla lip erect, lower almost none: stemless.
5. **M.** Kelloggii, Curran. Becoming a span or a ft. high: lower lip larger. Cal.

§ 3. **EUNANUS**, Gr. Style glandular: capsule not 1-sided.

6. **M.** Rattani, Gr. A span high: calyx very viscid. Mt. Tamalpais and Lake Co., Cal.
7. **M.** mephiticus, Greene. Corolla 6-8 lines long. Sierra Nevada.
8. **M.** nanus, H. & A. A span high or less, blossoming from near base. Cal., N.
9. **M.** Whitneyi, Gr. Dwarf, 1 or 2 in. high: corolla ½ in. long. Alpine, Cal.
10. **M.** Fremonti, Gr. Leaves narrow: corolla rarely white. Common in S. Cal.
11. **M.** subsecundus, Gr. Diffuse: flowers spicate, turned to one side. Cal.
12. **M.** leptaleus, Gr. A span or less high, often depauperate. Mts., Cal.
13. **M.** Torreyi, Gr. A span or more high, simple or branching. S. N. Mts.
14. **M.** Bolanderi, Gr. Very viscid, strong scented, 1-3 ft. high, simple. Cent. Cal.

15. **M. brevipes**, Benth. Very viscid, 1-2 ft. high: leaves slender. Monterey, S.

§ 4. **MIMULASTRUM**, Gr. Corolla throat contracted at mouth: limb rotate.
16. **M. pictus**, Gr. Simple stems or basal branches erect. Tehachapi, Cal.

§ 5. **EUMIMULUS**, Gr. Calyx plicately angled: style smooth; stigma-lobes flat.
17. **M. cardinalis**, Dougl. Viscid-villous, 2-4 ft. high: leaves ovate, cross.
18. **M. Lewisii**, Pursh. More slender, greener. Subalpine. Cal.-Or.
19. **M. primuloides**, Benth. Scapes 1-4 in. long: light green. S. N. Mts.
20. **M. moschatus**, Dougl. Stems 1-3 ft. long: leaves oblong-ovate, 1-2 in. long.
Var. longiflorus, Gr. Less viscid, corolla longer.
Var. sessilifolius, Gr. Leaves sessile: corolla 1 in. long. *M. inodorous*, Greene.
21. **M. moniliformis**, Greene. Leaves sparingly denticulate. S. N. Mts.
22. **M. dentatus**, Nutt. Simple stems a foot high or less. Humboldt Bay, N.
23. **M. luteus**, L. Erect, ½ to 4 ft. high: corolla large; palate prominent.
24. **M. Scouleri**, Hook. Erect, 1-2 ft. high: flowers smaller. Columbia R.
25. **M. glaucescens**, Greene. Corolla 1 in. long and broad, not dotted. S. N. Mts.
26. **M. nasutus**, Greene. Corolla short, often with a spot. Common.
27. **M. nudatus**, Curran. Corolla ½-¾ in. long, deep yellow. Cal.
28. **M. laciniatus**, Gr. Slender: ½-1 ft. high. Merced R., Cal.
29. **M. alsinoides**, Dougl. Slender, branching, 3-12 in. high. Moist rocks.
30. **M. inconspicuus**, Gr. Leaves ovate or narrower, entire, ½ in. long or less.
Var. acutidens, Gr. Calyx-teeth subulate: leaves denticulate. King's R.
Var. latidens, Gr. Calyx-teeth triangular ovate. Monte Diablo, S.
31. **M. Pulsiferæ**, Gr. Branching, 6-8 in. high: leaves 3-nerved. N. Cal. to W.
32. **M. peduncularis**, Dougl. Erect, slender: leaves ½-½ in. long. Columbia R.
33. **M. floribundus**, Dougl. Similar, slimy, musky. Common.
34. **M. Parishii**, Greene. Stout, erect, 1-2 ft. high, leafy. Los Angeles, S.
35. **M. bicolor**, Hartweg. Leaves small: calyx ribbed. Foot-hills, S. N. Mts.
36. **M. montioides**, Gr. Branching from base or simple: leaves slender. S. N. Mts.
37. **M. Suksdorfii**, Gr. Leaves often reddish, ½-½ in. long. Cascade and S. N. Nts.
38. **M. rubellus**, Gr. Leaves lanceolate, ½-1½ in. long. Cascade and S. N. Mts.
39. **M. Palmeri**, Gr. Leaves narrow: corolla-limb nearly rotate. S. E. Cal.
Var. androsaceus, Gr. *M. androsaceus*, Curran. Much branched: leaves broader:
corolla smaller, 3-6 lines long. S. E. Cal.

§ 6. **MIMULOIDES**, Gr. Calyx deeply cleft, almost nerveless.
40. **M. exilis**, Durand. *M. pilosus*, Watson. Much branched, leafy, very floriferous.
Common in Cal.

11. HERPESTIS, C. T. Gaertner.

1. H. rotundifolia, Pursh. Creeping in wet places: leaves obovate. Fresno, Cal.

12. GRATIOLA, Linnæus.

1. G. ebracteata, Benth. Leaves lanceolate: sepals equaling corolla.
2. G. Virginiana, L. More viscid: leaves broader: calyx much shorter.

13. ILYSANTHES, Rafinesque.

1. I. gratioloides, Benth. Diffuse: leaves ovate or oblong: corolla ½ in. long.

14. LIMOSELLA, Linnæus.

1. L. aquatica, L. Tufts 1-2 in. high: leaves fleshy, slender; semi-aquatic.

15. DIGITALIS, Linnæus.

1. D. purpurea, L. Tall stems with terminal spike of rose or white flowers mostly spotted. The common *Foxglove* now naturalized. Humboldt Bay to Or.

16. SYNTHYRIS, Bentham.

1. S. rotundifolia; Gr. Scapes naked 3-4 in. high, not exceeding leaves. Or.
Var. cordata, Gr. Leaves smaller, cordate, simply crenate. N. W. Cal.
2. S. reniformis, Benth. Scapes bracteate, surpassing leaves, pedicels shorter. Or.

17. VERONICA, Linnæus.

   * *Perennials subaquatic: racemes in the axils of opposite leaves: corolla blue.*
1. V. Anagallis, L. Leaves sessile oblong-lanceolate, subclasping.
2. V. Americana, Schweinitz. Leaves often petioled, broader. Common.
3. V. scutellata, L. Slender: leaves sessile, linear or lanceolate, acute.
   * * *Perennials: racemes terminal: leaves broad, an inch long or less.*
4. V. Cusickii, Gr. Stems erect 3-4 in. high, leafy: naked peduncle 3-9-flowered.
5. V. alpina, L. Not so leafy: racemes dense: corolla smaller, 2-3 lines broad.
6. V. serpyllifolia, L. Stems creeping and branching: spike-like raceme leafy.
   * * * *Lower annuals: flowers in the axils of mostly alternate leaves.*
7. V. peregrina, L. Nearly glabrous, erect, branching: flowers small.
8. V. arvensis, L. Pubescent, soon spreading: lower leaves crenate.
9. V. Buxbaumii, Tenore. Very pubescent decumbent; pedicels long. S. F. Bay

18. CASTILLEIA, Mutis.

Leaves and bracts all linear-lanceolate and entire: calyx all green. 1
Leaves mostly entire, narrow: calyx deeper cleft before than behind, mostly red: corolla 1½-2 in. long; upper lip (galea) ¾-1 in. long.................................... 2, 3

Calyx about equally cleft before and behind: floral bracts more or less dilated, red varying to yellow or whitish.

    Galea (upper lip) as long or longer than the tube, lip very short....... 4, 5, 6
    Galea hardly as long as tube: leaves linear, entire, white-woolly............. 7
    Galea much shorter than tube, about twice as long as the lip............. 8, 9

1. O. **stenantha,** Gr. Slender: corolla 1-1½ in. long. Cent. to S. Cal.
2. O. **affinis,** H. & A. Calyx cleft twice as deep before as behind. Cal. Coast.
3. O. **linearifolia,** Gr. Calyx cleft much deeper before. S. N. Mts.
4. O. **latifolia,** H. & A. Diffuse, viscid-villous: leaves broad, obtuse. Cal.
5. O. **parviflora,** Bong. Leaves laciniate-cleft or entire: galea partly exserted.
6. O. **miniata,** Dougl. Leaves and bracts mostly entire: galea more exserted.
7. O. **foliolosa,** H. & A. Many stems from woody base leaves crowded. Cal. Coast
8. O. **pallida,** Benth., var. occidentalis, Gr. Flowers whitish: low: alpine.
9. O. **Lemmoni,** Gr. Taller: spike dense, reddish. Sierra Co., Cal.

19. ORTHOCARPUS, Nuttall.

§ 1. **Castilleoides,** Gr. Lip of the corolla simply or somewhat triply saccate the lobes
(teeth) erect: anthers all 2-celled: bracts with colored tips.
Perennial: very leafy: leaves mostly 3-5-parted: galea obtuse........................ 1
Annual (as are all the remaining species): galea nearly straight............ ...... 2, 3, 4
Galea densely bearded, incurved at apex: filaments hairy........................... 5
§ 2. Corolla-lip simply saccate; teeth inconspicuous or wanting: galea ovate-triangular:
anthers 2-celled.
Bracts colored, dilated: corolla rose-purple...................................... 6, 7
Bracts not colored, leaf-like, 3-5-cleft, lobes acute.............................. 8, 9
§ 3. **Triphysaria,** Benth. Corolla-lip conspicuously 3-saccate, teeth minute or small;
tube slender: bracts all similar to the leaves.
Slender, diffusely spreading: minute flowers in nearly all the axils................ 10
Stems erect, often corymbosely or fastigiately branched.
    Stamens soon exserted: spikes of yellowish or white flowers dense........... 11
    Stamens included; anthers 1-celled: leaf divisions filiform.............. 12, 13
    Stamens included; anthers 2-celled but lower cell often imperfect........ 14, 15
    Stamens included; anthers 2-celled: stems strict, often simple.
        Very leafy and hirsute above: spike very dense; bracts broad.......... 16
        Spikes leafy: corolla yellow; sacs 2 lines long: viscid................ 17
            corolla white sacs 1-2 lines deep................. 18, 19

1. O. **pilosus,** Watson. Sierra Nevada above 5000 ft. to Mts. of Oregon.
2. O. **attenuatus,** Gr. Slender: spike of pale flowers very slender. Coast.
3. O. **densiflorus,** Benth. Spikes dense: leaves entire or few-lobed. Cal.
4. O. **castilleioides,** Benth. Spikes shorter: leaves mostly laciniate. Coast.

5. O. purpurascens, Benth. Crimson or rose-color spikes showy. Cal.
6. O. imbricatus, Torr. Slender: cor. hardly ½ inch long. S. N. and Cascade Mts.
7. O. pachystachyus, Gr. Low, stout: cor. over 1 inch long, galea hooked. N. Cal.
8. O. bracteosus, Benth. Hirsute, strict: corolla rose-purple. Br. Col. to Cal.
9. O. luteus, Nutt. Corolla golden yellow: galea obtuse straight. S. N. Mts.
10. O. pusillus, Benth. Leaves 3-5-parted into filiform lobes, often brownish.
11. O. floribundus, Benth. Erect 3-8 in. high: corolla ½ in. long. S. F. Bay.
12. O. erianthus, Benth. Corolla sulphur-yellow; galea dark. Cal. Coast.
Var. lævis, Gr. Often a foot high: corolla yellow to white: galea pale.
Var. roseus, Gr. Corolla larger, white or rose-color. San Francisco.
13. O. Bidwelliæ, Gr. Similar: smaller in every way. Sacramento Valley.
Var. micranthus, Gr. Still smaller: lip a line broad. Fresno Co., Cal.
14. O. gracilis, Benth. Bracts with purplish tips: corolla purplish. Rare.
15. O. campestris, Benth. 2-4 in. high: leaves mostly entire: corolla white.
16. O. lithospermoides, Benth. Corolla yellow or rose-tinged, large.
17. O. lacerus, Benth. Hairy leaves and bracts 3-7-cleft. S. N. Mts. and Sac. Val.
18. O. hispidus, Benth. Soft-hairy: spike slender. Or. and Cal.
19. O. linearilobus, Benth. Hirsute, stouter, more branched. Cent. Cal.

20. CORDYLANTHUS, Nuttall.

* *Calyx 2-leaved: flowers short-peduncled or sessile subtended by 2-4 bractlets: stamens 4: filaments hairy: corolla nearly included.*

Leaves mostly 3-5-parted, the upper and bracts hispid-ciliate...................... 1
Leaves entire except the 3-parted bracts: soft villous............................... 2
Leaves entire very slender; bracts obtusely 3-5-lobed, fan-shaped................... 3
Leaves entire very slender: flowers scattered on slender branches................... 4
* * *Calyx of one posterior leaf: flowers in short spikes, sessile in the axils of clasping bracts: no bractlets: low salt-marsh plants... 5, 6*
1. C. filifolius, Nutt. Corolla over ½ in. long, purplish. Cal.
2. C. pilosus, Gr. Tall: viscidulous flowers few in clusters or solitary.
Var. Bolanderi, Gr. Lower, more viscid: flowers all scattered.
3. C. Pringlei, Gr. Corolla 4-5 lines long, pale yellow. Clear Lake, Cal.
4. C. tenuis, Gr. Paniculate, 1-2 ft. high. Central Cal.
5. C. mollis, Gr. Stamens 2: anthers 2-celled. S. F. Bay.
6. C. maritimus, Nutt. Stamens 4. San Diego to Humboldt Bay.

21. PEDICULARIS, Tournefort.

Galea with a slender projecting or upturned beak: corolla dull rose or crimson: spike naked. Alpine in S. N. Mts... 1, 2
Galea with incurved beak: corolla white or whitish........................... 3, 4, 5
Galea falcate with subulate beak: a pair of stem leaves............................. 6

Galea not beaked: leaves pinnately-parted, lobes pinnatifid................... 7, 8, 9
1. P. Grœnlandica, Retz. Spike glabrous: corolla ¼ in. and beak ½ in. long.
2. P. attollens, Gr. Spike woolly; beak of galea 2-3 lines long.
3. P. contorta, Benth. Leaves pinnate, linear lobes incised. Or.
4. P. racemosa, Dougl. Leaves undivided, crenate: raceme leafy. Subalpine.
5. P. Howellii, Gr. Leaves entire, serrate or pinnate. Siskiyou Mts.
6. P. ornithorhyncha, Benth. Spike interrupted: calyx inflated. Mt. Tacoma.
7. P. palustris, L. (Var.) Calyx 2-cleft: corolla ½ in. long, purplish.
8. P. Semibarbata, Gr. Nearly stemless: spikes sessile. Mts. Cal.
9. P. densiflora, Benth. Corolla scarlet or crimson; galea ½ in long.

OROBANCHACEÆ.

1. APHYLLON, Mitchell.

Scapes or long peduncles from a scaly, fleshy rootstock or short stem............. 2, 1
Stems rising above ground: pedicels shorter than the flower or none.
　　Flowers an inch or more long on distinct pedicels........................ 3, 4
　　Flowers nearly sessile, about ½ in. long: anthers glabrous................. 4, 5
1. A. uniflorum, Gr. Scapes few: corolla often violet tinged: calyx lobes slender.
2. A. fasciculatum, Gr. Peduncles often many: corolla yellow: calyx lobes short.
3. A. comosum, Gr. Calyx-lobes half as long as pink or purple corolla.
4. A. Californicum, Gr. Calyx-lobes and bractlets nearly equaling corolla.
5. A. tuberosum, Gr. Stems thick, 1-3 in. high: flowers densely crowded.
6. A. pinetorum, Gr. More slender, ½-1 ft. high: flowers looser. Oregon.

2. BOSCHNIAKIA, C. A. Meyer.

1. B. strobilacea, Gr. A thick, brownish red spike of striped flowers.

LENTIBULARIACEÆ.

1. UTRICULARIA, Linnæus.

Stems stout, densely leafy: leaves 2-3-pinnate, very bladdery...................... 1
Stems filiform: leaves scattered, repeatedly forked, bristly...................... 2
Stems slender: leaves 2-ranked, not bladdery, forking............................ 3
1. U. vulgaris, L. Scapes 5-16-flowered: corolla ½ in. broad or more.
2. U. minor, L. Scapes 3-7 in. high: corolla 2-3 lines broad: spur short.
3. U. intermedia, Hayne. Scape 1-4-flowered: corolla ½ in. broad.

VERBENACEÆ.

Corolla nearly equally 5-lobed: calyx 5-toothed: small flowers in spikes............... 1
Corolla bilabiate, 4-lobed: calyx 2-cleft: small heads on slender peduncles....... 2

1. VERBENA, Tournefort.

1. V. officinalis, L. Spikes filiform, bracts minute; leaves pinnatifid.
2. V. polystachya, HBK. Leaves serrate: corolla a line broad. Rare.
3. V. hastata, L. Erect, 3-6 ft high: leaves coarsely serrate, petioled.
4. V. prostata, R. Br. Diffuse, spreading, hairy: corolla 2 lines broad.
5. V. bracteosa, Michx. Similar, but rigid bracts exceeding smaller flowers.

2. LIPPIA, Linnæus.

1. L. nodiflora, Michx. Creeping: peduncles 1-4 in. long: flowers rose to white.

LABIATÆ.

§ 1.   Stamens 4, coiled in the bud, much exserted through a cleft in the upper lip: leaves entire: corolla and curved filaments blue or purple............................. 1

§ 2.   Stamens erect or ascending, the posterior pair shorter or wanting: anther-cells short, close together or united: upper lip of corolla not concave or hooded (except in No. 9).

* *Corolla small, almost equally 4-lobed: axillary flowers in dense whorl-like clusters, the upper axils flowerless.*

Stamens 4, nearly equal, all perfect: calyx 5 toothed................................ 2
Stamens 2, with anthers, posterior pair sterile or wanting.......................... 3

* * *Corolla bilabiate: stamens 4.*

Flowers capitate: calyx equally 5-toothed: stamens distinct, straight.
    Upper lip of corolla entire or merely notched................................ 4
    Upper lip 2-cleft: stamens exserted....... 5
Flowers solitary or in clusters, axillary: curved stamens all perfect.
    Flowers small, white or purple: not ½ in. long............./................. 6
    Flowers over an inch long, orange; peduncles bracteate....................... 7
Flowers in oblong heads or interrupted spikes, blue or purple...................... 8
Flowers in axillary clusters, rose and white: upper anthers imperfect.............. 9
Flowers with white or purplish corolla an inch long: stamens perfect.............. 10

§ 3.   Stamens 2, the upper pair rudimentary or wanting: anthers 1-celled or with 2 cells widely separated on the ends of a filament-like connective: flowers in dense terminal heads or globose whorls (except sp. 8 of No. 1).
Connective versatile upon the short filament. Leaves pinnatifid.................. 11

Connective joined to the filament by one end; only one anther cell................................ 12
§ 4. Stamens 4, perfect: corolla bilabiate: calyx 15 nerved.
Flowers in oblong peduncled heads, pale violet: stamens exserted.................... 13
§ 5. Stamens 4, perfect, ascending under the concave or hooded upper lip.
Calyx with a projection on upper side: flowers solitary axillary.... 14
Calyx purple-tinged, upper teeth broad, obtuse, lower two lanceolate................ 15
Calyx 10-toothed, the shorter 5 teeth spiny, recurved: corolla small, white.......... 16
Calyx nearly equally 5-toothed: flowers in whorls or interrupted spikes............. 17

1. TRICHOSTEMA, Gronovius.

1. **T. oblongum**, Benth. Corolla-tube shorter than the calyx. Or.–Cal.
2. **T. laxum**, Gr. Diffuse: leaves petioled: cymes peduncled, often forked.
3. **T. lanceolatum**, Benth. Gray-green: leaves crowded, sessile. Or.–Cal.
4. **T. ovatum**, Curran. Leaves round-ovate: calyx densely villous. S. Cal.
5. **T. lanatum**, Benth. Shrubby: leaves narrow: corolla woolly. Santa Barbara, S.

2. MENTHA, Linnæus.

1. **M. Canadensis**, L. Villous: leaves oblong-ovate or narrower.
Var. glabrata, Benth. The similar serrate acute leaves nearly glabrous.

3. LYCOPUS, Tournefort.

1. **L. Virginicus**, L. Stem obtuse-angled: sterile filament minute. Or.
2. **L. lucidus**, Turcz, var. **Americanus**, Gr. Stem acute-angled, stout: runners tuberiferous: calyx-teeth slender, equaling corolla.
3. **L. sinuatus**, Ell. Leaves mostly incised or pinnatifid. N. Cal., Or.

4. PYCNANTHEMUM, Michaux.

1. **P. Californicum**, Torr. Leaves ovate or narrower, sessile, 1-3 in. long.

5. MONARDELLA, Bentham.

§ 1. Calyx over ½ in. long: corolla-tube much longer than the lobes.............. 1, 2
§ 2. Calyx ¼.½ in. long: corolla-tube but little longer than the linear or oblong flat lobes: flowers in dense involucrate heads.
　　　* *Tufted: corolla flesh-color to rose, lobes linear: calyx-teeth soft.*
Leaves ovate to lanceolate, petioled, pinnately veined: bracts obtuse.......... 3, 4, 5
Leaves linear to oblong, entire, ½-¾ in. long, upper subsessile.................... 6, 7
　* * *Annuals loosely branching: leaves entire or undulate, rather distinct, narrowed into a petiole: calyx-teeth with margined nerve.*
Bracts rigidly cuspidate white and transparent except the veins................. 8, 9

Bracts acute or obtuse nervose, less transparent or the outer green............. **10, 11**
Bracts broadly ovate, white-scarious, nervose with cross veins: corolla white or nearly so,
only 3 or 4 lines long: calyx-teeth with scarious tips....................... **12, 13**
1. **M. macrantha,** Gr. Orange red or scarlet corolla 1-1½ in. long. San Diego.
2. **M. nana,** Gr. Similar, more hairy: corolla white, rose-tinged, smaller. S. Cal.
3. **M. hypoleuca,** Gr. Densely white-tomentose: bracts nervose. S. E. Cal.
4. **M. villosa,** Benth. Soft-hairy or glabrate: bracts pinnately veined. Western Cal.
5. **M. odoratissima,** Benth. Nearly glabrous: bracts thin, whitish or purplish.
6. **M. linoides,** Gr. Ashy-pubescent: bracts scarious, white, pinkish. S. Cal.
7. **M. Palmeri,** Gr. Green: bracts very obtuse: otherwise like the last. S. Cal.
8. **M. Douglasii,** Benth. Bracts silvery between pinnate nerves and margin.
9. **M. Breweri,** Gr. Bracts broader, less translucent, wanting marginal nerve.
10. **M. lanceolata,** Gr. Bracts acute, cross veinlets between the nerves.
11. **M. undulata,** Benth. Bracts broadly ovate, not cross-veined. Coast.
12. **M. candicans,** Benth. Bracts with greenish nerves. Cent. Cal.
13. **M. leucocephala,** Gr. Bracts whiter, lightly nerved calyx-teeth slender.

6. MICROMERIA, Bentham.

1. **M. Douglasii,** Benth. Creeping: leaves round-ovate: pedicels slender. Coast.
2. **M. purpurea,** Gr. Erect: leaves lanceolate: flowers in dense clusters.

7. CALAMINTHA, Tournefort.

1. **C. mimuloides,** Benth. Hirsute, viscidulous. Monterey Bay.

8. POGOGYNE, Bentham.

Stamens all perfect: stigmas nearly equal: corolla ½-¾ in. long................. **1, 2, 3**
Stamens 2 perfect: stigmas very unequal: corolla ⅓ in. long....................... **4, 5**
1. **P. Douglasii,** Benth. Spikes oblong, white-hispid, bracts acute.
2. **P. parviflora,** Benth. Smaller; bracts mostly obtuse. S. F. Bay, N.
3. **P. nudiuscula,** Gr. Flowers in whorl-like clusters: bracts less hispid. S. Cal.
4. **P. ziziphoroides,** Benth. Flowers mostly in heads or short spikes.
5. **P. serpylloides,** Gr. Flowers in whorls or long interrupted spikes.

9. ACANTHOMINTHA, Gray.

1. **A. ilicifolia,** Gr. Rigid, 3-6 in. high: leaves broad, often cuspidate-toothed.
2. **A. lanceolata,** Curran. Taller: leaves lanceolate: flowers larger, an inch long.

10. SPHACELE, Bentham.

1. **S. calycina,** Benth. Shrubby: leaves rugose: hairy ring in corolla-base.

11. SALVIA, Linnæus.

1. C. carduacea, Benth. White-woolly, thistle-like: lavender corolla 1 in. long.
2. C. Columbariæ, Benth. Branching: leaves pinnatifid: blue corolla ½-¾ in. long.

12. AUDIBERTIA, Bentham.

Corolla 1½ in. long, crimson-purple: large leaves very rugose........................ 1
Corolla ½ in. long or less, violet or bluish.............................. 2, 3, 4, 7
Corolla ½-¾ in. long: stems woody below, 3-10 ft. high..................... 5, 6, 8
1. A. grandiflora, Benth. Stout, slightly woody. S. F. Bay, south.
2. A. incana, Benth. Leaves not rugose, 1 in. long or less. San Diego.
3. A. humilis, Benth. A span high, simple, stems nearly naked, base leafy.
4. A. stachyoides, Benth. Several ft. high: forming dense thickets. Cal. Coast.
5. A. Palmeri, Gr. Leaves oblanceolate, acute: whorls 4-8, distant. San Diego.
6. A. Clevelandi, Gr. Similar: leaves obtuse: whorls fewer: viscid. San Diego.
7. A. nivia, Benth. White-hoary, 3-4 ft. high: stamens exserted. Santa Barbara, S.
8. A. polystachya, Benth. Mostly very white: flowers in a thyrsus. S. Coast.

13. LOPHANTHUS, Benth.

1. L. urticifolius, Benth. Green, 4 to 6 ft. high: leaves ovate or cordate, large.

14. SCUTELLARIA, Linnæus.

Corolla slender. ½-¾ in. long, deep blue or violet: leaves ovate....................... 1
Corolla larger, ½-1 in. long, violet-blue: leaves oblong or narrow................ 2, 3
Corolla white or dull yellow: upper leaves entire, obtuse....................... 4, 5
1. S. tuberosa, Benth. Soft-hairy, mostly 3 or 4 in high: many tubers. Cent. Cal.
2. S. angustifolia, Pursh. Stems slender: corolla hairy inside, ¾-1 in. long.
3. S. antirrhinoides, Benth. Similar: leaves and corolla broader and shorter.
4. S. Californica, Gr. Slender: leaves short-petioled; upper short.
5. S. Bolanderi, Gr. More pubescent, very leafy: leaves sessile, broad, veiny.

15. BRUNELLA, Tournefort.

1. B. vulgaris, L. Simple stems ending in a dense spike of violet flowers.

16. MARRUBIUM, Tournefort.

1. M. vulgare, L. Hoary, bitter. Common Horehound naturalised.

17. STACHYS. Tournefort.

Corolla white or whitish: leaves soft-hairy or white tomentose................. 1, 2, 3
Corolla purple or rose, ½ in. long or less; tube not exceeding calyx................... 4
Corolla purple or rose; tube exceeding the calyx............................ 5, 6, 7

1. S. ajugoides, Benth. Softly white-hairy: leaves oblong obtuse: ill-scented.
2. S. albens, Gr. White-woolly, leafy, often tall: leaves rather acute.
3. S. pycnantha, Benth. Tawny-hairy, leafy: leaves obtuse, spike short, dense.
4. S. palustris, L. Leaves ovate-lanceolate, acute, mostly sessile. Or.
5. S. bullata, Benth. Mostly hispid, light rose flowers in interrupted spikes.
6. S. Chamissonis, Benth. Much larger: leaves 2-5 in. long. Wet ground.
7. S. ciliata, Dougl. Similar; leaves thinner, less hairy. Or. Coast.

PLANTAGINACEÆ.

1. PLANTAGO, Tournefort.

Stamens 4. Leaves not fleshy, 1-8-ribbed or nerved.......................... 1, 5, 6
     Leaves somewhat fleshy, oblanceolate to lanceolate or broader.... 2, 3, 7
     Leaves fleshy, linear to filiform; spike dense, cylindrical................ 4
Stamens 2: leaves linear to filiform, 1-4 in. long: annuals........................ 8, 9
1. P. major, L. var. Asiatica, Decne. Leaves ovate or oval: scape ½-2 high.
2. P. eriopoda, Torr. Yellowish wool at base: scape ½-1 ft. high. N. Cal. to Alaska.
3. P. macrocarpa, C. & S. Petioles long: capsule ¼-½ in. long. Coast, Wash.
4. P. maritima, L. Corolla-tube pubescent: seeds 2 to 4. Common on the Coast.
5. P. lanceolata, L. Petioles slender: scape deeply furrowed; spike short. Nat.
6. P. Patagonica, Jacq. Usually silky-woolly: slender leaves acute: spikes short.
7. P. hirtella, HBK. Scape with long dense spike, 1-2 ft. high. Cal. Coast.
8. P. Bigelovii, Gr. Spike dense, ½-1 in. long: capsule 4-seeded. Saline marshes.
9. P. heterophylla, Nutt. Spike 2-5 in. long, very slender: seeds 10-28. Cal.

CLASS II.—ENDOGENS OR MONOCOTYLEDONS.

ALISMACEÆ.

Flowers perfect: stamens usually 6: carpels in a whorl.
  Carpels numerous, distinct, obovate-oblong, flattened: scape paniculata........ 1
  Carpels 6-12, united at base, tapering to a beak: scape simple................. 2
Flowers monœcious or diœcious: carpels many, capitate, winged, short-beaked........ 3

1. ALISMA, Linnæus.

1. A. Plantago, L. Scape with branches in whorls: leaves ovate to lanceolate or narrower: petals small, white or pinkish. In water or mud.

2. DAMASONIUM, Jussieu.

1. D. Californicum, Torr. Scapes 6-18 in. high: leaves ovate to narrowly lanceolate, long-petioled: flowers in 3 or 4 whorls; pedicels 1-2 in. long: petals 3-4 lines long, incised above, white. In water or mud. S. N. Mts.

3. SAGITTARIA, Linnæus.

1. S. variabilis, Engelm. Leaves ovate-sagittate or some linear: flowers mostly in 3's: petals white, rounded $\frac{1}{2}$-$\frac{3}{4}$ in. long: tubers edible. In water or mud.

ORCHIDACEÆ.

Herbs with a more or less irregular perianth of 3 sepals and 3 petals; the lower petal (made so by a twist in the inferior ovary), called the *lip*, usually unlike the other two which generally resemble the sepals. Stamens and style united to form the *column* which is capped by a single perfect 2-celled anther, or (in *Cypripedium*) with a perfect anther on each side of the stigma over which curves a triangular sterile stamen. Our genera are usually grouped in four tribes here briefly defined.

I. Anther resting lid-like upon the column, deciduous: pollen-masses 4 .... 1, 2, 3
II. Anther united with column, persistent on its face above stigma: pollen-masses 2 .. 4
III. Anther erect on top of column, persistent: pollen-masses 2-4 .... 5 to 9
IV. Anthers 2, lateral; the sterile stamen petaloid, incurved...................... 10

* *Herbs with one to many green leaves: not parasitic.*

Leaf solitary from a globose corm. Scape 1-flowered................................. 1
                Scape 6-20-flowered............................ 3
Leaves several to many from a creeping rootstock.................................... 6
Leaves a pair below the raceme of small flowers, ovate or cordate.................... 7
Leaves 2 or 3 to many clasping slender or stoutish stems, at least at base.
     Flowers not leafy-bracted, white or greenish, in spikes or racemes.
         Lip of perianth spurred at base.. 4
         Lip not spurred: spike twisted: flowers 3-ranked............................ 5
     Flowers leafy-bracted, pedicellate, 1-20.
         Lip concave at base, constricted in middle................................. 8
         Lip an inflated sac, the mouth with incurved margin.......................10
       * * *Plants with no green leaves: stems simple, scape-like.*

Flowers and stems brownish, purplish or yellowish, often mottled or striped.......... 2
Flowers and stems nearly or quite white.. 9
Flowers and stems greenish: bracts membranaceous, acute. Sp. 8 in............... 4

1. CALYPSO, Salisbury.

1. C. borealis, Salisb. Stem 3-6 in. high: slender bract at top, subtending a drooping showy flower: sepals and petals lanceolate, rose-tinged ½-⅔ in. long; lip saccate, 2-spurred, mottled. Springy places or bogs, from Russian River (*Miss Wood*) to Br. Am. and E. to the Atlantic. Also in N. Eu. and Asia.

2. CORALLORHIZA, Haller.

Sepals and petals similar; lip dilated, recurved, flat or concave, 2-ridged at base. Column incurved. Rootstocks coral-like, hence the name *Coral-root.*

1. C. multiflora, Nutt. Sepals and petals 3-4 lines long, yellowish or whitish, purple tinged: spur formed by decurrent side-sepals wholly adnate to the ovary; lip broadly ovate, 3-lobed, the middle lobe with undulate or denticulate margin, often purple-mottled. Mts., San Diego to Br. Col., E. to the Atlantic.
2. C. Mertensiana, Bong. Similar: flowers red: lower half of the spur free: lip narrower, entire or with small teeth at base. Humboldt Bay to Alaska.
3. C. innata, R. Br. Smaller: sepals 1½-2 lines long: spur very short. Wash.
4. C. Bigelovii, Watson. Stout: sepals and petals oblong, obtuse, 4 lines long, purple-veined; spur none. S. N. & Coast Mts., Cal.
5. C. striata, Lindl. Similar: perianth 6-7 lines long. S. N. Mts., N. & E.

3. APLECTRUM, Torrey.

1. A. hiemale, Torr. Scape a foot high or more: leaf plaited, 4-8 in. long: perianth ½ in. long, greenish brown; lip whitish, somewhat spotted, deeply 3-lobed, 3-ridged. The glutinous bulbs give the name *Putty-root.* Or. E. to the Atlantic.

4. HABENARIA, Willd.

Stems slender: leaves few and at base: perianth 2 lines long or less............... 1, 2
Stems leafy. Spur 4-6 lines long, slender: lip narrow....................... 3, 4, 5
Spur short and thick.. 6, 7

1. H. elegans, Bolander. Spike dense: sepals and petals equal. Coast, Monterey, N.
2. H. Unalaschensis, Watson. Spike less dense: flowers smaller: bracts ovate.
3. H. leucostachys, Watson. Stout: flowers many, white: capsule sessile. Swamps.
4. H. sparsiflora, Watson. Lower, more slender: leaves narrower: greenish flowers 10-20, distant, exceeded by the slender bracts: capsule sessile. S. N. Mts. & N. Cal.
5. H. pedicellata, Watson. Raceme loose: capsule tapering into a pedicel.

6. **H. Cooperi**, Watson.   Stout: lip ovate: upper sepal ovate.   San Diego.
7. **H. gracilis**, Watson.   Like No. 4: lip linear: spur saccate.   Or., Wash.
8. **H. Michæli**, Greene.   Stout stem, leafless: spike dense: sepals ¼ in. long.   S. Cal.

5. SPIRANTHES, Richard.

1. **S. Romanzoffiana**, Chamisso.   Spike dense, conspicuously bracteate, 1-4 in. long:
    perianth greenish-white, ½ in. long, curved (Called *Ladies' Tresses*).   Wet places.
2. **S. porrifolia**, Lindl.   Similar: flowers smaller: 2 callosities at base of lip.

6. GOODYEARA, Robt. Brown.

1. **G. Menziesii**, Lindl.   Scape pubescent, 6-15 in. high: leaves smooth, 2-3 in. long,
    in a rosulate tuft: spike of puberulent white flowers 1-sided.   (*Rattlesnake Plantain*.)

7. LISTERA, Robt. Brown.

1. **L. Convalarioides**, Nutt.   Slender, 3-12 in. high: flowers purplish in a pubescent
    raceme: lip 2-lobed or emarginate, toothed at base, 2-5 lines long.   Damp woods.
2. **L. cordata**, R. Br.   Smaller: flowers minute, smooth.   (*Twayblade*.)   N. Cal., N.

8. EPIPACTIS, Haller.

1. **E. gigantea**, Dougl.   Leafy, 1-4 ft. high: leaves ovate to lanceolate: flowers 3-10,
    greenish, purple-veined: sepals ovate-lanceolate, ½-¾ in. long; lip as long.   Along
    streams.

9. CEPHALANTHERA, Richard.

1. **C. Oregana**, Reich. f.   Parasite: perianth ½ in. long; sepals and petals lanceolate;
    lip as in *Epipactis* with wavy-crested nerves.   Forests, N. Cal to Or.

10. CYPRIPEDIUM, Linnæus.

1. **C. fasciculatum**, Kellogg.   Villous, 2-6 in. high: leaves ovate, a pair: peduncle
    viscid: flowers several in a cluster or 1, greenish.   (*Bradley's Cypripedium*).   Rare.
2. **C. montanum**, Dougl.   Leafy, 1-2 ft. high: flowers 1-3; sepals and wavy-twisted
    petals brownish, narrow, 1½-2½ in. long: lip oblong, white, purple-veined.   Cent.
    Cal. to Or.
3. **C. Californicum**, Gr.   Often taller: flowers 3-12: sepals ½ in. long: lip obovoid-
    globose.   In swamps or wet places.   N. Cal.

IRIDACEÆ.

Perennial herbs with sword-shaped or grass-like leaves, the divisions of the superior
perianth all petaloid and convulute in the bud, withering-persistent.

12

Outer segments of perianth larger, recurved or spreading; the inner erect or incurved;
style-branches petaloid, curving over the linear anthers.......................... 1
Segments nearly alike: stigmas filiform: filaments often united...................... 2

1. IRIS, Tournefort.

* *Perianth-tube stem-like above the ovary, ½-3 in. long: stems leafy*.................. 1, 2
* * *Perianth-tube short and funnelform above the ovary.*
Stems leafy: bracts (enclosing peduncles and buds) green, often distant............ 3, 4
Stems naked or with 1 or 2 leaves, terete: floral bracts not distant................ 5, 6
Stems with many bracts and rigid radical leaves.................................. 7
Stems with 2-3 short bract-like leaves, 2-flowered................................ 8

1. I. **macrosiphon**, Torr.   Stems very slender, flattened, surpassed by the dark
   green grass-like leaves: flowers rich purple-blue, on short pedicels; tube 1-3 in. long;
   sepals 1½-2 in. long.   S. F. Bay to Humboldt Bay.   Placer Co.
2. I. **Douglasiana**, Herbert.   Stouter and taller stems: leaves and bracts broader,
   pedicels longer; tube shorter; sepals usually with a white center, blue-purple or lilac;
   often yellow or buff.   S. F. Bay to Siskiyou Mts.
3. I. **Hartwegi**, Baker.   Stems slender, flattened, 2-9 in. high: leaves 2-3 lines wide:
   flowers light colored.   S. N. Mts.   June.
4. I. **tenax**, Dougl.   Similar, taller, 1-flowered: flowers larger, bright lilac-purple,
   segments 2-2½ in. long.   Or. to Br. Col.
5. I. **longipetala**, Herbert.   Stems stout, equaling the leaves, 3-5-flowered: sepals
   lilac or whitish, purple and yellow veined, 2½-3 in. long.   Monterey to Or.
6. I. **Missouriensis**, Nutt.   More slender: leaves narrower: bracts dilated, scarious,
   1-1½ in. long: flowers pale blue.   N. Cal. to Or.
7. . I. **bracteata**, Watson.   Leaves striate, sides unlike: perianth yellow, 2-3 in. long.
   (*Howell's Iris.*)   Discovered by Thos. Howell in S. W. Or., 1884.
8. I. **tenuis**, Wats.   Perianth white, veined with yellow and purple, 1½ in. long.
   (*Henderson's Iris.*)   Discovered by L. F. Henderson in Or., 1881.

2. SISYRINCHIUM, Linnæus.

1. S. **bellum**, Wats.   ·Flowers blue, purple-striped, ½-1 in. broad.   Cal., Or.
2. S. **Californicum**, Ait. f.   Scape winged: flowers yellow.   Coast.   Wet places.
3. S. **grandiflorum**, Dougl.   Flowers red-purple, 1-1½ in. broad.   N. Cal. to Br. Col.

LILIACEÆ.

§ 1.   Floral bracts not leaf-like: perianth persistent: anthers introrse: style entire.
   * *Flowers in umbels or heads upon naked scapes: root a bulb or corm.*
   a.   Perianth parted to the base or nearly so: stamens at base; anthers versatile.

Flowers rose-purple to white: bracts broad: odor of onions............................. 1
Flowers greenish white: bracts narrow: slender leaves several......................... 2
Flowers yellow: pedicels jointed at top: leaves one or several........................ 3
b. Perianth not parted to the base: stamens on the throat.
Perianth-tube thin, somewhat inflated and angular or saccate: stamens on the throat in
   one row; anthers basifixed, 3 alternating with petaloid staminodia or smaller anthers:
   ovary nearly or quite sessile... 4
Perianth-tube thicker, opaque, not inflated or saccate: anthers basifixed, 3 alternating
   with petaloid staminodia: filaments decurrent to base: ovary sessile............. 5
Perianth-tube not inflated or saccate: filaments in two rows (except sp. 11); anthers
   versatile.. 6
Perianth-tube subcylindrical, 6-saccate at base, scarlet or crimson, the short segments
   yellowish: stamens 3 alternating with broad staminodia....................... 7
    * * *Flowers on short scape-like pedicels, umbellate on an underground peduncle.*
Perianth salver-form; tube slender, 1-2 in. long; lobes half as long................... 8
* * * *Flowers in racemes or panicles; perianth segments distinct and anthers versatile*
                                (except No. 12.)
a. Stems scape-like or sparingly leafy, arising with many leaves from a bulb.
Flowers blue or white, ½-1½ in. long, slightly one-sided, in a simple raceme........... 9
Flowers white or whitish, 2-5 lines long in a dense nearly simple raceme............ 10
Flowers white or pinkish, scattered on branches: withering perianth twisted........ 11
Flowers white or yellowish, paniculate: perianth-tube equaling reflexed lobes....... 12
b. Stems not scape-like, simple: rootstock slender: white flowers small.
   Leaves 2-ranked, sessile, often clasping, lanceolate to ovate.................... 13
   Leaves 2, rarely 3, petioled, cordate: perianth-segments 4: stamens 4........... 14
c. Stems rigid: lower bracts and rigid leaves spine-tipped: flowers 1-3 in. long...... 15
§ 2. Floral bracts none or leaf-like: perianth segments distinct, deciduous: anthers
   extrorse or opening on the sides.   In No. 24 the perianth is persistent: anthers
   introrse.
a. Stem simple, from a scaly bulb: leaves often whorled: perianth segments similar:
   anthers versatile: style long: fruit a capsule: seeds flat, horizontal.
Segments oblanceolate, with a groove: style entire: stigma large, 3-lobed........... 16
Segments broader, not groved: style entire or 3-cleft; stigmas small................ 17
b. Stem from a coated corm: anthers basifixed.
Leaves a pair at the base, broad: perianth-segments similar, lanceolate, recurved.... 18
Leaves few, linear-lanceolate: perianth-segments unlike, the inner (petals) broad.... 19
c. Stem branching, leafy above: rootstock slender: flowers nodding or hanging.
Flowers apparently axillary: anthers 1-2-awned or pointed above, sagittate.......... 20
Flowers white or greenish, terminal, in clusters or solitary, beneath the leaves..... 21
d. Stem a scape or scape-like from a rootstock: large leaves basal: flowers umbellate or
   solitary, red or white: filaments hairy: ovary 2-celled: fruit a many-seeded berry. 22

a. Stemless: leaves a pair, broad; flowers umbellate on an underground peduncle: pedicels 3-cornered prostrate and curved in fruit: stamens 3: styles 3, divergent.... 23
f. Stem with 3 broad leaves at top and a single flower: outer segments green...... 24
§ 3. Bracts greenish or scarious: flowers in a simple raceme or panicle: segments distinct, persistent: anthers small: styles or sessile stigmas persistent; capsule deeply
3-lobed.
a. Stem tall, leafy: leaves large ovate to lanceolate, nerved, plicate: panicles large. 25
b. Stem from a coated bulb, leafy at base: leaves linear or grass-like, smooth.
Flowers white, erect: yellow glands at base of segments.........:................. 26
Flowers yellowish or purplish, nodding glandless................................ 27
c. Stem equitant-leafy, from a rootstock: leaves slender; anthers 2-celled, introrse.
Flowers small, greenish, each with a cup-like or 3-lobed involucre.................. 28
Flowers yellowish-green: filaments woolly; style none.......................... 29
d. Stem with a large tuft of grass-like stiff leaves from a rootstock: raceme of white
flowers very dense, long: anthers extrorse: styles reflexed or coiled............ 30

1. ALLIUM, Linnæus.

§ 1. Bulbs connected with rootstocks: leaves 2 or more: capsule not conspicuously
crested.
Scape round, 1-2 ft. high, exceeding the leaves: bracts 2, large, acuminate: bulb white.. 1
Scape flattened above: umbels often nodding: stamens and style slender; bracts united a
base 2, 3
§ 2. Bulbs without rootstocks: scape not flattened, slender: leaves very slender.
a. Leaves 2 or more, shorter than, or scarcely exceeding the scape.
Ovary obscurely crested: perianth rose: stamens included: scape 3-10 in. high... 4, 5, 6
Ovary distinctly 6-crested.
Perianth-segments white or light pink becoming thin and lax.
Bracts 2, short, acute; stamens included................................ 8 9
Perianth rose-color: filaments deltoid-widened at base............... 10, 11
Filaments filiform crests conspicuous 12, 13
b. Leaves 2 or more, much exceeding the very short scape................... 14, 15
§ 3. Scape much flattened, 2-edged, short: leaves 2, linear, flat, falcate: flowers rose-
color.
Bracts 2; stamens included... 16, 17, 18
Bracts 3-5: stamens not included: leaves ½-1 in. broad........................ 19
· 1. A. unifolium, Kell. Leaves 2-4: segments 5-7 lines long exceeding stamens.
2. A. validum, Wats. Scape 1-3 ft. high: bracts 2-4, broad: pedicels ¼-½ in. long:
segments slender 3-4 lines long: bulb-coats white. Alpine, July to Sept.
3. A. hæmatochiton, Wats. Scape slender, 4-12 in. high: bracts 2, short: flowers
deep purple or rose-color: bulb-coats deep reddish purple, shining. S. Cal. Coast.

4. **A. acuminatum**, Hook. Perianth-segments serrulate, 4-7 lines long, tips acuminate, recurved, rigid in fruit. Washington to Cent. Cal. Rare.
5. **A. Bolanderi**, Wats. Similar: flowers rarely white: stamens adnate to the middle, half as long as the segments which are nearly straight. N. W. Cal.
6. **A. lacunosum**, Wats. Scape 3-6 in. high: pedicels ¼-½ in. long: stamens nearly equaling perianth; filaments a little expanded at base. Cent. Cal.
7. **A. Sanbornii**, Wood. Slender, 1-2 ft. high: perianth 2-3 lines long. S. N. Mts.
8. **A. attenuifolium**, Kell. Leaves filiform, sheathing the scape near base.
9. **A. hyalinum**, Curran. Perianth thin, transparent in fruit: capsule 1-seeded.
10. **A. serratum**, Wats. Outer bulb-coats with distinct zigzag lines along which they tear horizontally into serrate strips: inner perianth segments shorter, narrower.
11. **A. bisceptrum**, Wats. Scapes often in 2's, rarely angular. S. N. Mts.
12. **A. campanulatum**, Wats. Flowers many: perianth light rose-color. S. N. Mts.
13. **A. Bidwelliæ**, Wats. Smaller: flowers fewer, smaller, bright rose. S. N. Mts.
14. **A. tribracteatum**, Torr. Scarcely 2 in. high; bracts 3. Mostly alpine.
15. **A. parvum**, Kell. Similar: bracts 2, shorter. Sierra Valley.
16. **A. falcifolium**, H. & A. Scape 2-5 in. high: capsule 3-crested. Coast Mts.
17. **A. Breweri**, Wats. Scape 1-3 in. high: crests 3, slightly lobed. Coast Mts.
18. **A. Lemmoni**, Wats. Taller leaves nearly straight. Sierra Valley.
19. **A. platycaule**, Wats. Scape and leaves broader. Montane. S. N. Mts.

2. MUILLA, Watson.

1. **M. maritima**, Wats. Perianth-segments 2-3 lines long, subrotate. Coast.

3. BLOOMERIA, Kellogg.

1. **B. aurea**, Kell. Scape 6-18 in. high: leaf solitary: each filament surrounded at base by a 2-cuspidate appendage: Coast Ranges, Monterey to San Diego.
2. **B. montana**, Greene. Larger: flowers an inch broad: cusps of the filament-appendage half as long as the filament: anthers 1½ lines long. S. Cal.
3. **B. Clevelandi**, Watson. Leaves several, very slender: style short. San Diego.

4. BRODIÆA, Smith.

[The next two genera are united with this in the Botany of California and the Cal. Flora. E. L. Greene of the University of California has recently elaborated the species under the generic names here given.]
Stamens 3, alternating with bifid or entire staminodia.................. 1, 2, 3
Stamens 6, 3 with petaloid appendages back of the anther.................. 4, 5
1. **B. volubis**, Baker. Twining scape 4-10 ft. high: perianth rose-color to white: sagittate anthers 2-appendaged on the back. *Stropholirion Californicum* Torr. Cent. Cal.

2. B. multiflora, Benth. Scape 2-4 ft. high: perianth violet-purple 8-10 lines long: staminodia obtuse, entire. Or. to Cent. Cal., June, July.

3. B. congesta, Smith. Scape 2-5 ft. high: purple staminodia bifid. B. C. to C. Cal.

4. B. pulchella, Greene. Perianth-tube, like the last, constricted above: distinguished by the stamens and strictly umbellate inflorescence. Cal., May, June.

5. B. capitata, Benth. Scape 6-18 in. high: bracts often dark purple: perianth-tube not constricted above. Very abundant in Cent. Cal., S. & E., Jan. to Apr.

5. HOOKERA, Salisbury.

1. H. Californica, Greene. Scape 2 ft. high: pedicels 2-3 in. long: perianth 1½-2 in. long, deep purple to rose-color: anthers ⅓ in. long, a little exceeded by the retuse staminodia. This and next under *Brodiæa grandiflora* in Cal. Bot. Sacramento Val. Much less common than the next species.

2. H. coronaria, Salisb. Smaller: anthers exceeding the acute staminodia.

3. H. minor, Britten. Scape 3-6 in. high: perianth-segments rotate: anthers 2 lines long exceeded by the emarginate or retuse staminodia. Sac. Val., S.

4. H. terrestris, Britten. Scape usually not rising above ground: pedicels 3-4 in. long: staminodia yellowish, margins involute. S. F. Bay, N.

5. H. stellaris, Greene. Scape 2-6 in. high: perianth red-purple: anthers 2-appendaged: staminodia longer, white. (*Purdy's Hookera*.) Near Ukiah.

6. H. rosea, Greene. Similar: perianth rose-red: stamens not appendaged; filaments triangular. Lake Co., Cal. Discovered by Mrs. Curran, May, 1884.

7. H. Orcuttii, Greene. Scape a foot or more high: staminodia wanting or obscure. San Diego. Discovered by C. R. Orcutt in 1884.

6. TRITELEIA, Douglas.

Perianth-tube broad at base: upper and inner stamens with winged filaments...... 1, 2
Perianth-tube tapering to a narrow base: filaments not winged or appendaged.
    Stamens in 2 rows: flowers not yellow................................... 3, 4, 5
    Stamens in 1 row; filaments broadening downward........................... 6
    Stamens in 2 rows or nearly equal: flowers yellow......................... 7, 8
Perianth-tube short; segments rotate, yellow: filaments with appendages........ 9, 10
Perianth open-campanulate, cleft below the middle: stamens in 1 row.......... 11, 12

1. T. grandiflora, Lindl. Pedicels ½-1 in. long, numerous: perianth light blue, 1 in. long: lower anthers sessile, upper on filaments which are winged below. Or. and Wash. E.

2. T. Howellii, Greene. Similar: upper filaments winged above. Or. & Wash.

3. T. candida, Greene. Scape 2-4 ft. high: perianth 1½ in. long, white: filaments coiled. Discovered by J. R. Scupham. Fresno Co., Cal., June 1886.

4. T. laxa, Benth. Umbel of usually 15-30 purple-blue flowers: anthers acute. Cal

6. **T. peduncularis,** Lindl.   Pedicels often 6-10 in. long: perianth rose-purple to nearly white, cleft below the middle, 1 in. long: anthers retuse.   Wet places Cent. Cal.

6. **T. Bridgesii,** Greene.   About a root high: umbel rather few-flowered: perianth light blue.   Very common in open forests about Humboldt Bay.   Chico.

7. **T. crocea,** Greene.   Perianth 7-9 lines long: lower filaments very short.   N. Cal.

8. **T. gracilis,** Greene.   Smaller: filaments subequal: anthers acute.   S. N. Mts.

9. **T. ixioides,** Greene.   Scape ½-2 ft. high: filaments unequal, wing-dilated, 2-appendaged above.   S. Cal. to Or.

10. **T. lugens,** Greene.   Similar: perianth dark brown outside: winged filaments not forked above.   Collected by E. L. Greene near Vacaville, Cal., May 4, 1886.

11. **T. hyacinthina,** Greene.   Perianth white with green veins, rarely purple-tinged: filaments broad at base, united into a ring.   Moist ground.   Cent. Cal. N.

12. **T. lilacina,** Greene.   Smaller: perianth lilac-purple: filaments not so broad at base, distinct.   Col. in Amador Co. by Mrs. Curran.   May, 25, 1886.

N.B.—No. 1 is *Brodiæa Douglasii*, Wats.; No. 11, *B. lactea*, Wats.; No. 2, 4, 5, 6, 7, 8, 9, have the same specific names under *Brodiæa* in the Cal. Bot. No. 3, 10, 12 are new. *Behria tenuiflora*, Greene, of the California Peninsula, is the type of a new genus belonging between this and the next. It is appropriately dedicated to Dr. H. H. Behr, of the University of California.

7. BREVOORTIA, Wood.

1. B. coccinea, Wats.   Flowers pendulous, 1-1¼ in. long.   N. Cal.   (*Firecrackers.*)

8. LEUCOCRINUM, Nuttall.

1. L. montanum, Nutt.   White flowers surpassed by the leaves.   Cal. E.

9. CAMASSIA, Lindley.

1. C. esculenta, Lindl.   Flowers irregular, lower segment deflexed: segments not connivent in age, persistent: seeds shining.   N. Cal., N & E to Montana.

2. C. Leichtlinii, Watson.   Nearly regular flowers larger; segments broader, connivent and twisted, at length deciduous: seeds obovoid, dull. S. F. Bay, N. to Wash.

10. HASTINGSIA, Watson.

1. H. alba, Wats.   Flowers in dense close raceme, 2-3 lines long.   N. Cal., Or.

2. H. bracteosa, Wats.   Flowers 3-6 lines long, nearly equaled by bracts: stamens short.   Coll. by Thos. Howell in Curry Co., Or., May, 1884.

11. CHLOROGALUM, Kunth.

Perianth-segments very slender, ⅜-⅝ in. long: pedicels longer than the bracts 1
Perianth-segments oblong-oblanceolate, ¼-½ in. long: bulb-coats, not fibrous 2, 3

1. **C. pomeridianum**, Kunth.   Bulbs densely fibrous: leaves crispate-undulate, mostly radical: flowers white, purplish-veined.   Cal.   (*Soap-root.*)
2. **C. parviflorum**, Watson.   Leaves grass-like: flowers pinkish.   San Diego.
3. **C. angustifolium**, Kellogg.   Leaves not undulate: white flowers.   N. Cal.

12.   ODONTOSTOMUM, Torrey.

1. **O. Hartwegi**, Torr.   Numerous flowers 4-6 lines long.   S. N. Foot-hills.   Rare.

13.   SMILACINA, Desfontaines.

1. **S. amplexicaulis**, Nutt.   Panicle close: segments and filaments similar: fragrant.
2. **S. sessilifolia**, Nutt.   Simple zigzag raceme few-flowered: berries blue-black.

14.   MAIANTHEMUM, Weber.

1. **M. bifolium, DC.   Var.** (?)   Zigzag stem 3-12 in. high.   S. F. Bay to Alaska.

15.   YUCCA, Linnæus.

1. **Y. baccata**, Torr.   Leaf-margins thread-bearing: perianth campanulate.   S. Cal.
2. **Y. Whipplei**, Torr.   Leaf-margins serrulate: perianth rotate spreading.   S. Cal.

16.   LILIUM, Linnæus.

Flowers horizontal to erect, spotless or finely dotted, white, purplish or pale yellow; segments tapering into long narrow claws, spreading.

    Flowers becoming purple or purplish: bulb-scales not jointed.............. 1, 2
    Flowers pale yellow, 3 in. long or more: bulb-scales jointed. 3
Flowers orange-yellow to red, spotted; segments oblanceolate to lanceolate.
    Flowers erect or horizontal, less than 2 in. long....................... 4, 5, 6
    Flowers nodding, segments revolute (*Tiger Lilies*)...................... 7, 8, 9

1. **L. Washingtonianum**, Kellogg.   Bulbs becoming 6-8 in. long, the scales thin, lanceolate, 2-3 in. long; stems 2-5 ft. high: leaves in several whorls (some scattered), ⅓-1 in. broad, undulate: flowers white becoming purplish, often dotted, horizontal on erect pedicels; segments 3-4 in. long, ½-¾ in. wide: yellow anthers 5-6 lines long.
2. **L. rubescens**, Watson.   Similar: bulb smaller, thicker, broader scales an inch long: stems 1-7 ft. high: flowers nearly white to lilac, becoming rose-purple, 1½-2 in. long: anthers 2-3 lines long.   Coast Mts., S. F. Bay to Klamath R.
3. **L. Parryi**, Wats.   Stem 2-5 ft. high: leaves mostly scattered, slender.   S. Cal.
4. **L. parvum**, Kell.   Flowers few to many, erect or nearly so: anthers 1-2 lines long: capsule sub-spherical, ½-¾ in. long.   S. N. Mts.   4–8,000 ft. alt., N. to Or.
5. **L. maritimum**, Kell.   Flowers horizontal, deep reddish orange.   S. F. to Hum'dt.
6. **L. Bolanderi**, Wats.   Stems 1-2-flowered: leaves mostly in whorls, 1-2 in. long: flowers nearly horizontal, brownish or dull purple.   Hum'dt to S. W. Or.

7. **L. pardalinum, Kell.** Rootstocks thick and branching, forming mat-like masses of bulbs: stems 3-7 ft. high: perianth segments 2-3 in. long, bright orange red with large purple spots below: anthers red, 4-5 lines long. Cent. Cal. to Or.

Var. **angustifolium, Kell.** Slender, small: leaves 3-4 lines broad, scattered.

8. **L. Humboldtii, Roezl & Leichtlin.** Bulbs 2-6 in. thick, often purplish, the fleshy ovate-lanceolate acute scales 2-3 in. long: stems purplish, 4-8 ft. high: leaves undulate in 4-6 whorls of 10-20 each: pedicels mostly 3-6 in. long: perianth-segments 3-4 in. long, ½-1 wide, papillose-ridged near base: anthers red, ½-⅔ in. long. Cal.

9. **L. Columbianum, Hanson.** Perianth-segments 1½-2 in. long: yellow anthers 2-3 lines long. Wash. to Cent. Cal.

17. FRITILLARIA, Linnæus.

Styles distinct above; stigmas linear: capsule obtusely angled.................. 1, 2, 8

capsule acutely angled or winged........ 4, 5, 6

Styles united: stigma 3-lobed: flowers not spotted: stamens unequal................ 7

1. **F. recurva, Benth.** Segments narrow, scarlet and yellow, spotted. Cal., Or.
2. **F. liliacea, Lindl.** Leaves near base: flowers greenish white. San Francisco Bay.
3. **F. biflora, Lindl.** Leaves near base: flowers dark brown, purple, green-tinged; segments widely spreading: mucronate anthers 2 lines long. Coast, San Diego to Mendocino.
4. **F. lanceolata, Pursh.** Bulbs with a few large scales and many like rice grains: leaves in 1-3 whorls: flowers dark purple mottled with greenish yellow; segments not spreading.

Var. **floribunda, Benth.** Flowers 1-8, lighter colored, blotched with brownish purple; segments acute, ⅛-⅜ in. broad, finely crenulate.

Var. **gracilis, Wats.** Flowers smaller with narrower acuminate segments.

5. **F. parviflora, Torr.** Flowers 3-20, with spreading segments ½-⅜ in. long, lighter colored than the last. Cent. S. N. Mts.
6. **F. atropurpurea, Nutt.** Capsule not winged, acutely 6-angled. S. N. Mts.
7. **F. pluriflora, Torr.** Stems leafy: flowers reddish purple, ⅔-1 in. long. Cent. Cal.

18. ERYTHRONIUM, Linnæus.

1. **E. grandiflorum, Pursh.** Leaves not mottled: flowers 1-6 or more, yellow, or cream color with darker center; segments recurved 1-2 in. long. Wash. to N. Cal.

Var. **Smithii, Hook.** Large flowers purple-tinged. Cent. Cal. Coast.

2. **E. Hartwegi, Watson.** Bulb ½-⅜ in. long: leaves mottled: flowers 1-3 on scape-like pedicels, light yellow and orange; segments scarcely recurved. S. N. Mts.
3. **E. purpurascens, Watson.** Leaves undulate: peduncle racemosely or subumbellately 4-8-flowered or more; pedicels very unequal: flowers light yellow, purple-tinged, orange center. S. N. Mts.

19. CALOCHORTUS, Pursh.

§ 1. Pedicels recurved in fruit: capsule broadly 3-winged.

Flowers on branching stems, nodding: concave petals closely connivent, hairy within, ciliate.................... .... 1, 2

Flowers on rather weak stems, erect or nearly so: fruit nodding or not stiffly erect.

Flowers yellow, 6-8 lines long, densely hairy within.............................. 3

Flowers white to lilac or blue. Petals covered with hairs..................... 4 to 7

Petals hairy below only, or naked........... 8, 9, 10

§ 2. Flowers and fruit erect on stout pedicels: capsules not winged (except in 11 & 12): petals and sepals often with spots. (*Mariposas* or *Butterfly Tulips*.)

Flowers lilac or purplish, 1-1½ in. long: capsules 3-winged..................... 11, 12

Flowers yellow, more or less marked with brown or purple.................. 13 to 16

Flowers white or lilac... 17 to 20

1. C. albus, Dougl. Petals white: sepals green, not spreading. Cal.

2. C. pulchellus, Dougl. Petals yellow or orange: sepals yellow or greenish, spreading. Coast Mts., Monterey to Mendocino.

3. C. Benthami, Baker. Slender, 3-6 in. high: leaves longer: anthers acute. S. N. Mts.

4. C. Maweanus, Leichtlin. Stem flexuose: petals covered above with white or blue-purple hairs, acute: anthers acuminate. N. Cent. Cal.

5. C. cæruleus, Wats. Very slender, 3-6 in. high: flowers 2-5 in an umbel: petals lilac dotted or lined with darker blue: anthers oblong, obtuse: capsule nearly orbicular. S. N. Mts.

6. C. elegans, Pursh. Similar: petals greenish white, scarcely ciliate: anthers long acuminate. Var. nanus, Wood, has acute more hairy petals, smaller. N. Cal. N.

7. C. Tolmiei, H. & A. Stouter, about a foot high: petals ⅔-1¼ in. long, lilac-tinged: anthers lanceolate, acuminate. Mt. Shasta to Or.

8. C. nudus, Wats. Flowers 1-6, usually in an umbel, white or lilac: sepals about equaling the broadly fan-shaped hairless petals: anthers obtuse. Cent. S. N. Mts.

9. C. lilacinus, Kellogg. Leaves rather broad: flowers 3-10, on long zigzag pedicels: petals ½-1 in. long, pale lilac, slightly hairy below: anthers much shorter than the filaments, obtuse. S. F. Bay, Geysers.

10. C. uniflorus, H. & A. Similar: flowers 1 or 2: gland densely hairy. W. Cal.

11. C. Greenei, Wats. Stout, 1-2 ft. high: sepals with yellowish hairy spot: petals densely yellow-hairy below: anthers ½ in. long. N. Cal. to Or.

12. C. Lyoni, Wats. Sepals naked: anthers 1½-2 lines long. Los Angeles.

13. C. clavatus, Wats. Petals covered with club-shaped hairs at base: gland orbicular, deep: anthers purple, 4-5 lines long, obtuse. S. Cal. Coast.

14. C. Weedii, Wood. Stem zigzag: petals deep yellow, dotted, covered with slender hairs: gland small, densely hairy: anthers mostly acute. Cal. Coast.

Var. purpurascens, Wats. Petals purple or purple-blotched. St. Barbara.

15. O. Obispoensis, Lemmon. Sepals longer than the rotate or recurved, long-hairy, often bifid petals. San Luis Obispo.

16. O. luteus, Dougl. Petals 1-2 in. long, from yellow to deep orange, with more or less brownish purple inside: gland broad, rounded or somewhat crescent-shaped, densely hairy: anthers yellow, obtuse. Very variable. San Diego to Mendocino and S. N. Mts.

Var. oculatus, Wats. Petals white lilac or yellowish with a dark central spot: gland usually a narrow crescent.

Var. citrinus, Wats. Petals deep or lemon yellow with central spot.

17. O. venustus, Benth. Like the last: petals white or pale lilac above, with a reddish spot near the top, a brownish spot in the center bordered with yellow and a brownish base: gland large, oblong, hairy. Monte Diablo, S.

Var. purpurascens, Wats. Deep lilac or purple form. Kern Co.

18. O. splendens, Dougl. Like the preceding: petals clear lilac, paler in center, claw darker: anthers purple, ¼-½ in. long. Monterey, S.

19. O. macrocarpus, Dougl. Sepals about equaling the obovate acute or acuminate purple-lilac petals, 1½-2 in. long: anthers ½-⅔ in. long. N. Cal. N.

20. O. Nuttallii, T. & G. Slender: a single stem-leaf, or rarely 2 or 3: petals cuneato-obovate, usually white above, with a purplish band above the yellow base, sometimes deep lilac. S. N. Mts.

20. STREPTOPUS, Michaux.

1. S. amplexifolius, DC. Stem 2-3 ft. high: peduncles twisted beneath the deeply cordate clasping leaves, usually forked or kneed: perianth greenish white, ¼-½ in. long, recurved above: anthers tapering into a single awn. N. Cal. N.

2. S. roseus, Michx. Smaller: flowers rose-purple: anthers 2-pointed. Or., N.

21. PROSARTES, D. Don.

Style slightly 3-cleft: fruit triangular, ½ in. long, bright salmon-color 1
Style entire: fruit ovoid or obovoid: leaves mostly cordate and clasping.
    Filaments longer than the anthers 2, 3, 4
    Filaments much shorter than the nearly sessile anthers 5

1. P. Menziesii, Don. Perianth-segments ½-1 in. long, acute. S. F. Bay, N.

2. P. Hookeri, Torr. Stamens nearly equaling or a little exceeding the perianth. ⅓-½ in. long: ovary hairy: style exserted. Russian Riv. to Monterey.

3. P. trachyandra, Torr. Similar: stamens shorter: ovary smooth. S. N. Mts.

4. P. Oregana, Wats. Flowers often purplish-veined: stamens exserted. Or.

5. P. parvifolia, Wats. Woolly: leaves 1-1½ in. long. Siskiyou Mts.

22. CLINTONIA, Rafinesque

1. **C. uniflora**, Kunth. Nearly stemless: peduncle shorter than the leaves, 1-2-flow-ered: perianth white, ⅔-1 in. long, pubescent. S. N. Mts. and Humboldt Bay, N.
2. **C. Andrewsiana**, Torr. Flowers rose-red in a globose umbel on a stout peduncle, often one or more smaller clusters below: fruit rich blue. In the redwoods.

23. SCOLIOPUS, Torrey.

1. **S. Bigelovii**, Torr. Perianth ½-¾ in. long: sepals lanceolate, spreading, striped: petals erect, very slender, dark: style branches 2-3 lines long. Redwoods.
2. **S. Halli**, Wats. Smaller: style-branches a line long. Cascade Mts.

24. TRILLIUM, Linnæus.

Flower sessile. Leaves sessile or nearly so, large.. 1
              Leaves long-petioled: stem 3-4 in. high............................ 2
Flower pedunoled. Leaves sessile or nearly so, rhombic-ovate....................... 3
              Leaves on petioles 1-15 lines long, lanceolate................... 4

1. **T. sessile**, L., var. Californicum, Wats. Very variable: petals lurid-purple or rose-red to white, 1-4 in. long. San Diego to Or.
2. **T. petiolatum**, Pursh. Petioles exceeding or equaling the blade. Or. & Wash.
3. **T. ovatum**, Pursh. Flowers white becoming rose, fragrant. Santa Cruz, N.
4. **T. rivale**, Wats. Slender: leaves 1-2 in. long. N. W. Cal. & S. W. Or.

25. VERATRUM, Tournefort.

Perianth-segments entire or serrulate, oblanceolate, thickened on the sides at base.. 1, 2
Perianth-segments fringed rhombic-ovate, the riged base divided by a narrow furrow.. 3

1. **V. Californicum**, Durand. Stout, 3-7 ft. high: leaves L-12 in. long, sheathing: panicle 1-2 ft. long: perianth-segments whitish with a greener brown-edged base.
2. **V. viride**, Ait. Flowers green in slender panicles. Oregon, N.
3. **V. fimbriatum**, Gr. Leaves lanceolate, 6-18 in. long, narrowed at base. Cal.

26. ZYGADENUS, Michaux.

Flowers all perfect; segments longer than the stamens, the outer ones not clawed..... 1
Flowers smaller; stamens equaling or exceeding the perianth, 2 or 3 lines long..... 2, 3

1. **Z. Fremonti**, Torr. From a few inches to 3 or 4 ft. high: raceme simple or compound: perianth rotate; segments ¼-⅜ in. long, rather obtuse. San Diego to Humb't.
2. **Z. venenosus**, Wats. Leaves rarely over 2 or 3 lines broad, usually folded: raceme simple or nearly so. Cent. Cal., N. (*Death Camass.*)
3. **Z. paniculatus**, Wats. Similar, stouter: raceme compound: lower flowers often sterile, short pedicelled; segments 2 lines long, triangular, acute. Cal., E.

27. STENANTHIUM, Gray.

L. **S. occidentale,** Gr. Slender, 1-2 ft. high: perianth 4-7 lines long; segments linear-lanceolate, tips recurved: linear seeds winged. Or., N.

28. TOLFIELDIA, Hudson.

L. **T. occidentalis,** Wats. Viscid-pubescent: involucre 3-lobed often reddish.
2. **T. glutinosa,** Willd. Involucre scarcely lobed, near the flower. Or., N. & E.

29. NARTHECIUM, Moehring.

L. **N. Californicum,** Baker. Raceme loose, 3-5 in. long: perianth 3-4 lines long; capsule bright salmon-color; seeds with tails at both ends. N. Cal.

30. XEROPHYLLUM, Michaux.

L. **X. tenax,** Nutt. Stem 2-5 ft. high: perianth-segments ½-¾ in. long. Cal., N.
2. **X. Douglasii,** Wats. Smaller in every way. Columbia River.

GLOSSARY OF SPECIFIC NAMES AND BOTANICAL TERMS.

All the specific names found in this work are here defined except a few of obscure or unknown meaning, and some which have undoubtedly been overlooked. Commemorative names are followed by the names—when known to me—of those thus honored. Specific names are given sometimes in one gender, sometimes in another. The learner must know that, as a rule, if a specific name ends in *us*, *a*, or *um*, it may end in either of the other two to correspond with the gender of the generic name; as, *Convolvulus Californicus* (Masculine), *Polygala Californica* (Feminine), *Galium Californicum* (Neuter.) Or, the specific name may end in *is* or *e*, the former agreeing with masculine and feminine generic names, the latter with neuter names. The meaning of each name, where possible, is given in a form suitable for a common or English name of the plant. Botanical terms are given in italic letters. Figures in parentheses indicate the number of times the name is used in this book.

Abortion, imperfect growth, or failure of an organ.
Abrotanifolia, abrotanus-leaved.
Abrupt, suddenly ending.
Acaulescent, stemless above ground.
Acerose, needle-shaped.
Achilleæfolia, yarrow-leaved.
Acicularis, *Acicular,* slender-acerose
Acuminata, *Acuminate,* tapering to a point.
Acuta, *Acute,* angle at the apex, less than 90 and greater than 25 degrees.
Acutangula, acute-angled.
Adenocaulon, glandular-stemmed
Adenophylla, glandular-leaved.
Adnate, adherent from the first (adnate anthers adhere to the filament by the back side.)
Adsurgens (2), upward-turning.
Adventitious, not in the usual place.
Affinis (5), closely related to other species.

Aggregata, aggregated.
Agrestis, field.
Ajugoides, ajuga-like; like bugle,
Akene, a seed-like fruit.
Alata, winged.
Alba (5), **Albens,** white.
Albescens, whitish, whitening.
Albicaulis (3), white-stemmed.
Albidus, whitish.
Albiflora (2), white-flowered.
Alismæfolia, alisma-leaved.
Alnifolia, alder-leaved.
Alpestre (2), mountain, growing on high mountains.
Alpina (2), *Alpine,* on the summits of lofty mountains.
Alsinanthemum, old generic name.
Alsinoides, alsine-like, like sandwort.
Alternate, one after another, not opposite.
Altissima, highest, growing on mountains higher than other species.
Alyssoides, alyssum-like.

(191)

Amarella, old generic name.
Ambigua (3), doubtful, too much like other species.
Americana, (6), American.
Amictum, clothed, covered.
Amœna (2), charming.
Amplectans, twining.
Amplexicaule (2), stem-embraced (by leaves.)
Amplexifolius, embracing leaves.
Anagallis, old generic name.
Anagalloides, anagallis-like, like pimpernel.
Andersoni, Dr. C. L. Anderson of Santa Cruz, who has specially studied seaweeds and willows.
Andrewsiana, **Andrewsii**, Dr. T. L. Andrews, Monterey, 1850.
Andromedia (meaning not significant.)
Androsacea (2), like androsace.
Androsæmifolia, androsæmon - leaved, leaves like St. John's wort.
Androus, in composition, means stamens; diandrous, or 2-androus, meaning stamens two.
Anglica (2), English.
Angustata, slender or narrow.
Angustifolia (7), narrow-leaved.
Annua, annual.
Anomala, anomalous, peculiar.
Anserina, old generic name of gooseweed.
Anterior, next to the observer, not toward the stem.
Anthylloides, anthylla-like, like musk-ivy.
Antirrhinoides (2), snapdragon-like.
Antiselli, Dr. Thos. Antisell.
Aparine, old generic name.
Apendiculata, appendaged.
Apetalous, without petals.
Aphanoptera, wingless.
Aphylla, leafless.
Apiculate, having a short abrupt point.
Appressed, lying or pressed close, as leaves to branches.
Aquatica, *aquatic*, living in water.
Arborea (3), tree-like.
Arbutifolia, arbutus-leaved.
Arcuata (4), curved, or jaundiced, *i. e.*, yellowish.

Arenaria, sand, growing in sand.
Argophylla, spotted leaf.
Arguta, aggressive.
Ariæfolius, aria-leaved.
Arida, dry, growing in dry places.
Aristatus, awned, bearded.
Aristella, small-awned.
Armeria, generic name.
Arnottii, Dr. Arnott, 1830-40.
Aromatica, aromatic.
Arvense (11), field.
Asarifolia, asarum-leaved.
Ascending, rising obliquely upward.
Asper (Aspera, Asperum) (2), rough.
Asperima, very rough.
Asplenifolia, asplenium leaved.
Asprella, rough.
Assurgentiflora, upward-turning flowers
Atractyloides, thistle like.
Atropurpurea (2), dark or black-purple.
Attenuata (2), *attenuate*, very slender and tapering.
Attenuifolium, attenuate-leaved.
Attollens, high-growing.
Aurea (5), golden.
Auriculate, **Aurita**, eared; bearing projecting lobes at the base.
Austinæ (4), Mrs. R. M. Austin, of Sierra Co.
Aquatilis, **Aquatica**, *Aquatic*, living in water.
Aquifolium, holly-leaved.
Axil, between the base of a leaf and the stem.
Azureus, azure, blue.

Baccata, berry-like.
Baileyi, W. W. Bailey.
Barbiger (Barbigera, Barbigerum) (3), bearded.
Bartsiæfolia, Bartsia-leaved.
Beckwithii (2), Lieut. E. G. Beckwith who commanded a Government expedition.
Bellidifolia, daisy-leaved.
Bellum (2), beautiful.
Benthami (2), Geo. Bentham, a great English botanist.
Bernardina, San Bernardino Co.
Betulæfolia, birch-leaved.
Biceptrum, two-stemmed (wands).

Bicolor (6), two-colored.
Bicornuta, two-horned.
Bidwelliæ (2), Mrs. Bidwell of Chico.
Biennis (2), *Biennial,* living two seasons.
Bifidum, *Bifid;* that is, cut to the middle.
Biflora (3), two flowered.
Bifolium (2), two-leaved.
Bigelovii (4), Dr. J. M. Bigelow.
Bilabiate, two-lipped.
Biloba, two-lobed.
Binghamæ, Mrs. R. F. Bingham.
Bioletti (2), F. T. Bioletti.
Bipinnate, twice pinnate.
Blade, the broad upper part of a petal.
Blanda, bland, pleasant.
Blattaria, the old generic name.
Blepharophylla, eyelash-leaved.
Blochmanæ, Mrs. Ida Blockman.
Bloomeri, H. G. Bloomer. (See Index.)
Bolanderi (15), Henry N. Bolander. See Index.)
Boreale (6), northern or boreal.
Botrys, an old generic name.
Bottæ, P. E. Botta, a French collector.
Brachy calyx, short calyx.
Brachyloba, short-lobed.
Brachycarpum (2), short-podded.
Brachysperma, short-seeded.
Bract, Bractlet, reduced leaves of a flower cluster.
Bracteata (2), Bracteosa (4), bracted.
Breviflorus, short-flowered.
Brevipes, short-peduncled.
Brevistyla (2), short-styled.
Breweri (16), W. H. Brewer, of Yale, who was chief of the botanical department of the California State Geological Survey, 1860-4.
Bridgesii (3), Thos. Bridges, who botanized mostly in South America.
Brownii, Brown, an English botanist.
Bryophora, moss-bearing or mossy.
Bulbifera, *Bulbiferous,* producing bulbs.
Bullata, blistered.
Bursa-pastoris, shepherd's purse.
Buxbaumii, J. C. Buxbaum, a German botanist.
Buxifolia, boxwood-leaved.

Caducous, falling soon, as the calyx of a poppy.

Cærulea (3) cerulean, dark-blue.
Cæsium, bluish-gray.
Cæspitosa (4), *Cespitose,* growing in tufts.
Californica (78), California.
Callicarpa, beautiful pods.
Calycina, pertaining to the calyx (large in this species).
Calycosa (4), large-calyx.
Campanularia, bell-bearing.
Campanulata (2), *Campanulate,* bell-shaped.
Campestre (5), field, growing in level fields.
Candicans, whitening, becoming white.
Candida, pure white.
Canadense (4), Canada.
Canescens (3), *Canescent,* gray, hoary.
Canina, dog.
Cannabinum, hemp.
Canus, ash-colored.
Capillaris (2), *Capillary,* hair-like.
Capitata (3), *Capitate,* in a head (the flowers).
Capparideum, caper.
Capsule, a dry fruit (pod) of more than one carpel.
Cardinale (2), chief, principal.
Carduacea, thistle-like.
Carneum, flesh-like.
Carnosula (2), flesh-colored.
Caroliniana (2), Carolina.
Carpel, one of the leaves forming a pistil.
Caseana, E. L. Case.
Cataria, cat.
Castilleioides, castilleia-like.
Caudata, *Caudate,* tailed.
Caudex, an erect rootstock or a stem rising but little above the ground.
Caulescens, *Caulescent,* stem-producing, having a stem.
Cauline, on the stem, as cauline leaves, not radical.
Centranthifolius, centranthus-leaved.
Cerasiformis, cherry-like.
Cereus, waxen.
Chamissonis, Adelbert von Chamisso, a German poet and botanist who, with Eschscholtz, Choris, an artist, and their commander, Kotzebue, visited San Francisco, Oct. 1816.

Cheiranthifolius, wall-flower leaved.
Childii, H. S. Child.
Chilensis, Chile.
Chloranthus, green-flowered.
Chorisiana, Choris (see Chamissonis.)
Chrysantha, yellow-flowered.
Chrysanthemifolius, chrysanthemum-leaved.
Cicutarius, cicuta-like.
Ciliata, *Ciliate,* fringed with parallel hairs.
Ciliosa, ciliate.
Cineria, ashy-gray.
Circinata (2), *Circinate,* coiled downward.
Circinatiformis, circular.
Citrinus (3), lemon-yellow.
Clavata (2), *Clavate,* club-shaped.
Claw, the slender basal part of some petals.
Cleft, cut about half way down.
Cleistogama, *Cleistogamous,* having flowers which do not open, but are fertilized in the bud.
Clevelandi (7), D. Cleveland, San Diego.
Coccinea (4), scarlet.
Coerulea, cerulean, blue.
Cohesion, the union of similar organs.
Collina (3), hill, growing on hills.
Collinsioides, collinsia-like.
Coloratum, colored.
Columbariæ, dove (doves eat the seeds of this Salvia).
Columbianum (2), Columbia
Columbinum (2), dove-like (color).
Commune, common.
Comosa (2), *Comose,* bearing tufts of hairs.
Concinna (3), beautiful.
Concolor (3), of one color.
Confertum (3), dense, crowded together.
Confluent, running together, joined.
Congdoni, J. W. Congdon, Mariposa.
Congesta (8), congested, bunched.
Conjugialis, conjugal; the fruit in pairs.
Connate, joined together; as opposite leaves.
Connective, that which joins anther cells.

Connivent, coming together.
Contorta, twisted.
Convalarioides, convalaria-like, like lily-of-the-valley.
Convolute, rolled up. (In flower buds, one edge of a leaf in and the other out all around).
Cooperæ, Mrs. Elwood Cooper, Santa Barbara.
Cooperi (2), Dr. J. G. Cooper, a noted California zoologist.
Cordata (3), *Cordate,* heart-shaped (leaves).
Cordifolia (4), cordate leaves.
Coriaceous, leathery.
Corniculata, small-horned.
Cornuta, horned.
Coronaria, crowned.
Corrugata, corrugate.
Corymbosa (2), *Corymbose,* like a *Corymb* which is a flat topped inflorescence; the lower branches as tall as the upper ones and the main stem.
Costate, ribbed.
Cotulæfolia, cotula-leaved, like mayweed leaves.
Cotyledon, a generic name.
Coulteri, Coulteriana, Dr. Thos. Coulter, who collected on this coast 1831-3.
Crassifolia (4), coarse or thick leaved.
Creeping, lying on the ground and rooting.
Crenate, edges with rounded teeth.
Cretica, Cretan.
Crinita, bearded.
Crispa, *Crispate.*
Crista-galli, cock's-comb.
Crocea (2), saffron-yellow.
Crotalariæ, Crotalaria, rattle.
Crystalina, crystaline.
Cucullaria, hood-like.
Cucullata (2), hooded.
Cuneata (3), *Cuneate,* wedge-like (leaves).
Cuneifolia (2), wedge-leaved.
Cupuliferum, cup-bearing.
Curassavicum, old generic name.
Curvisiliqua, curved-pod.
Curtipes, short-pediceled.
Cusickii (2), W. C. Cusick.
Cuspidate, armed with a cusp.

Cylindrica, cylindrical.
Cymbalaria (2), old generic name.
Cyme, a flat topped cluster with the oldest flower in the center.
Cyathiferum, cup-bearing.
Cytisoides, cytisus-like, like golden-chain.

Davisæ, Miss N. J. Davis.
Debilis (2), weak, tender.
Deciduous, leafless in winter.
Declined, bending downward.
Decora (3), pretty.
Decumbens (2), Decumbent.
Deflexa, Deflexed.
Dehiscence, the opening of a pod or anther which is
Dehiscent, i.e. does not remain closed as does a pea-nut.
Deltoidea, triangular.
Demissa, low, dwarfed.
Densiflora (6), dense-flowering.
Densifolia, densely-leaved.
Dentata (3), Dentate, toothed with erect teeth.
Denticulata (3), Denticulate, finely toothed.
Depauperata (2), small as if not well nourished.
Depressa, Depressed, pressed down.
Deustus, burnt (application not obvious).
Diadelphous, stamens in two sets.
Dianthoides, dianthus-like, pink-like.
Dichlamydeum, having both calyx and corolla.
Dichotoma (3), Dichotomous, forking, stems and branches dividing.
Dictyota (2), netted (leaves).
Didyma. Didymous, double (the fruit).
Didymocarpus, double-pod.
Diffusa (7), Diffuse, loosely spreading.
Digitate, palmate, leaflets on the end of petiole.
Digynum, two carpeled, or two-styled.
Dilatum, spreading, broad.
Diœcious, bearing staminate and pistillate flowers on separate plants.
Dipetala, two-petaled.
Diploscypha, double-cupped.

Discolor (2), two-colored. (The sides of the leaf unlike in color).
Dispersa, dispersed, scattered.
Dissecta (2), Dissected, cut in many lobes.
Distans, standing apart.
Distichum, two-ranked.
Distinct, not united.
Divaricata (7), Divaricate, separating widely.
Divergent, separating.
Diversifolius (2), variously leaved.
Diversiloba, variously lobed.
Douglasii (15), Douglasiana (2), David Douglas, a Scotch botanist, who collected in 1825 and 1831-2-3.
Draba, a generic name.
Drummondii, Mr. Drummond, who botanized on the plains and this coast before 1840.
Drupe, fruit like a plum or cherry.
Drymarioides, drymaria-like.
Dumetorium, of the thickets.
Dumosa (4) bushy.

Ebracteata, bractless.
Echinata, spiny, like a hedge-hog.
Ecornuta, hornless.
Edulis, edible.
Eiseni (2), Dr. Gustav Eisen, naturalist and viticulturalist.
Elæaginifolia (2), elæaginus-leaved.
Elatum, tall.
Elegans (5), elegant.
Elliptica (4), Elliptical, in the form of an ellipse twice as long as broad.
Emarginata, Emarginate.
Eminens, high, tall ; or growing in high places.
Emoryi, Major W. H. Emory, who commanded an exploring expedition in 1846.
Empetriformis, empetrum-like, like (crow-berry).
Engelmanni, Dr. Geo. Engelmann, St. Louis, author of many monographs upon difficult genera.
Epigynous, growing upon the pistil.
Epilobioides, epilobium-like, resembling willow-herb.
Equilaterale, equal-sided (the leaves).

Erianthus. woolly-flower.
Eriocarpa, woolly-pod.
Eriocephalum, woolly-head.
Eriopoda, woolly feet (the base of the plant).
Eriophorus, wool-bearing.
Erosa, *Erose,* ragged-edged as though gnawed.
Esculenta, esculent, edible.
Europæa, European.
Eupatoria, old generic name.
Eurycarpa, broad-podded.
Exigua (3), dwarf, small.
Exilis (2), slender, feeble.
Exserted, projecting beyond the other organs.
Extipulate, without stipules.
Extrorse. facing outward.

Falcifolium, falchion-leaved.
Farinosa, starchy.
Fascicle, a close bunch of rather long-stemmed flowers of equal height.
Fasciculata (4), fascicled, fascicle-bearing.
Fastigiate, close, parallel and erect branches.
Ferruginea, rusty.
Fertile, fruitful. (Fertile anthers produce pollen).
Ficus-Indica, Indian-fig (the old name).
Filament, the stem of a stamen, a thread.
Filicaulis, thread-like stems.
Filiform, thread-like.
Filifolia (2), thread-leaved.
Filipes, thread-stemmed flower (the pedicel thread-like).
Fimbriata (3), *Fimbriate,* fringed.
Fimbriolate, bearing a fringe.
Flaccida. flaccid.
Flammula, the old generic name.
Flavescens (2), turning yellow.
Flavulum, yellowish.
Flexuose, bending in a zigzag way.
Floccosa, *Floccose,* bearing tufts of woolly hairs.
Floribunda (7), many-flowered.
Foliacea, *Foliaceous,* leafy.
Follicle, a simple pod opening along the ventral suture only.
Foliolate, pertaining to leaves (3-foliolate, having 3 leaflets, etc.)

Foliolosa (3), leafy.
Fontana, fountain, growing around springs.
Formicissimus, beautifully-formed.
Formosa (4), beautiful (in form).
Fragarioides, strawberry, like fragaria.
Fragrans, fragrant.
Franciscana (2), San Francisco.
Franklinii, Franklin, an early botanist.
Free, not adherent to other organs.
Fremonti (6), Gen. John C. Fremont.
Frigidum, frigid, growing in cold places.
Froebelli, Julius Fræbel, 1855.
Frondosa, leafy.
Fruit, the ripened pistil and all that adheres to it.
Fruticosa (2), *Fruticose,* shrubby.
Fucatum, painted.
Fugacious, soon disappearing.
Fulcratus, spurred.
Fullonum, fuller's (used by fullers in dressing cloth).
Fusca, dusky, dark-colored.
Fusiform, spindle-shaped.

Gallica, Gallic, French.
Galioides, galium-like.
Gambelii (2), Dr. Wm. Gambel, an ornithologist.
Gauræfolia, gaura-leaved.
Gelida, frost-loving.
Gibbous, swollen out, sack-like.
Gibbsii, G. W. Gibbs.
Gigantea (3), gigantic.
Gilioides, gilia-like.
Githago, old generic name.
Glabella (4), nearly smooth or hairless.
Glaber (Glabra, Glabrum) (5), hairless.
Glaberima, very smooth or hairless.
Glabrata, *Glabrate,* becoming smooth in age.
Glandulosa (5), glandular.
Glauca (6), *Glaucous,* covered with a white powder.
Glaucescens (2), bluish-gray, slightly glaucous.
Glechoma, generic name.
Glechomæfolia, glechoma-leaved.
Glomerata, densely-clustered.

Glomerule, a dense head-like cluster of cymules.
Glutinosa (3), glutinous.
Goodrichii
Gordoni, Gordonianus, Gordon, an English botanist.
Gracile (9), slender.
Graciliflora, slender-flowered.
Gracilenta (3), slender-growing.
Grande (2), grand.
Grandiflora (15), large-flowering.
Grandifolia, large-leaved.
Gratissima, most-pleasing.
Grayi (5), Dr. Asa Gray, the greatest American botanist of the 19th Century.
Greenei (4), Edward L. Greene, Professor of Botany in the Catholic University of America.
Greggii, Dr. Gregg.
Grisea, bluish-gray.
Groenlandica, Greenland.
Gratioloides, gratiola-like.
Guttatus, spotted.
Gymnocarpa, naked-fruited.
Gypsophiloides, gypsopila-like.

Hæmatochiton, red-coated (the bulbs.)
Hallii (4), E. Hall, who collected in Oregon, in 1871.
Hanseni, Geo. Hansen.
Harknessii, Dr. H. W. Harkness, a student of fungi.
Hartwegi (6), Theodore Hartweg, a German who collected in Cal. in 1846-7.
Hassei, Dr. Hasse.
Hastata, *Hastate*, spear-shaped.
Hebecarpus, blunt-podded.
Hederacea, ivy-like.
Heermannii, Dr. A. L. Heermann, who collected in Sacramento Val., 1853-6.
Hendersoni (4), L. F. Henderson, an Oregon teacher and botanist.
Hesperium (2), evening or western
Heterantha, variable-flowered.
Heterodon, variously-toothed.
Heterodoxa (2), heterodox, unlike others of its kind.
Heterophylla (6), variously-leaved.

Hexandra, six-anthered.
Hickmani (2), J. B. Hickman, Monterey.
Hiemale, winter.
Hippurioides, hippuris-like.
Hirsuta (4), *Hirsute*, hairy, with rather coarse hairs.
Hirsutissima (2), most-hirsute.
Hirta, shaggy, rough.
Hirtella (3), roughish-hairy.
Hispida (5), *Hispid*, bristly, with stiff hairs.
Hispidissima, most-hispid.
Hispidula (3), slightly-hispid.
Humilis (5), low.
Humistrata (2), spreading on the ground.
Humboldtii, Baron von Humboldt, the great naturalist.
Holboellii
Holosericens, wholly-silky.
Hookeri (5), Hookerianus, Sir Joseph Hooker.
Hornemanni.
Hornii, Dr. Geo. H. Horn, U. S. Army, 1862-4.
Horrida, horrid.
Howellii (9), Thos. Howell (see index).
Howardi, W. J. Howard.
Hyacinthina, hyacinth-like.
Hyalinum, thin, membranous (the perianth).
Hydrophylloides, like Hydrophyllum, which means water-leaf.
Hypogynous, growing below the pistil; i. e., upon the receptacle.
Hymenosepalus, skinny-sepaled.
Hypoleuca, white below (the leaves).
Hypopitis, generic name.
Hyssopifolia, hyssop-leaved.

Ida-Maia, (see page 75).
Ilicifolia (3), holly-leaved.
Imbricata (2), *Imbricate*, overlapping like shingles.
Incana (3), hoary, ashy-gray.
Incarnata, flesh-colored, fleshy.
Incisa, *Incise*, *Incisely*, cut into sharp lobes with sharp sinuses.
Incompta, plain, unadorned.
Inconspicua (2), inconspicuous.

Inferior, below (an inferior ovary is covered by the adherent calyx).
Inflata, inflated.
Innata, *Innate*, with the anthers attached by the base.
Insignis (2), marked, noticeable.
Integerrimus, perfectly entire.
Integrifolia (2), entire-leaved.
Intermedia (3), intermediate.
Interrupta, broken between (the spike with naked spaces).
Intertexta, interwoven.
Introrse, facing in (anthers).
Involucrata (2), *Involucrate*, having an *Involucre*, a whorl of calyx-like leaves.
Involucel, a little involucre.
Involute, rolled inward.
Irregular, organs of a set (particularly the petals) unlike.
Ixioides, ixia-like.

Jalapa
Jaffrayanus
Jamesii, Dr. Edwin James.
Jaredi, Lorenzo Jared, San Luis Obispo.
Jasminoides, Jasmine-like.
Jeffreyi, John Jeffrey, a Scotch gardener.
Jepsoni, Willis L. Jepson, California State University.
Jocundum, **Jocundus** (3), pleasing.
Jonesii, Marcus E. Jones, who has specially studied the flora of Utah and adjacent regions.
Juliflora
Junceum, rush-like.

Kamtschatica, Kamtschatka.
Keel, the lower pair of petals in a leguminous flower.
Kelloggii, Dr. Albert Kellogg. (See Index.)
Kingi (2), Clarence King, geologist and author.
Kleei, W. C. Klee.
Kentrophyta, generic name.

Lacerus, lacerated, torn.
Laciflora, torn-flower.
Laciniata (6), *Laciniate*, slashed, cut into narrow lobes.

Lacteum, milky white.
Lacunosum, full of holes or hollows.
Lacustre, Lake.
Lætus, pleasing, agreeable.
Læve (3), smooth.
Lævicaulis, smooth-stemmed.
Lamina, blade, as of a leaf.
Lanatum, woolly.
Lanceolata (5), *Lanceolate*, lance-shape.
Lasiantha, hairy-flowered.
Lasiocarpum, hairy-podded.
Lasiococcus, hairy berry.
Lasiophyllum, hairy-leaved.
Lasiosperma, hairy-seeded.
Lateral, on the side of a stem.
Lateriflora, lateral-flowered.
Latidens, broad-toothed.
Latifolia, broad-leaved.
Latipes, broad, *i. e.*, flat-pediceled.
Laurina, laurel-like.
Laxa (3), loose.
Laxiflorus, loose-flowered.
Leana, L. W. Lee, Oregon.
Ledebourii
Ledifolius, ledum-leaved.
Legume, a pod like that of pease.
Leichtlinii
Leiocarpa, smooth-podded.
Lemmoni (11), J.G. Lemmon. (See Index.)
Lepida (2), charming.
Lepidota, scaly, covered with scurf.
Leptalea (2), very slender.
Leptandra, slender anther.
Leptantha, slender-flowered.
Leptocarpa, slender-podded.
Leptophyllus, slender-leaved.
Leptosepala, slender-sepaled.
Leptostachya, slender-spiked.
Leucantha (2), white-flowered.
Leucocephala (2), white-headed.
Leucodermis, white-skinned.
Leucophæus, ash-colored.
Leucophylla (2), white-leaved.
Leucopsis, white.
Leucostachys, white-spiked.
Lewisi (3), M. Lewis. (See index.)
Ligusticifolia, ligusticum-leaved.
Liliacea, **Lilicina**, lily-like.
Limb, the border or spreading part of a calyx or corolla of united leaves.

Limonium, generic name.
Limosa, swamp.
Lindheimeri (2), Dr. F. Lindheimer.
Lindleyi, John Lindley, an eminent English botanist, 1799-1865.
Line, a twelfth of an inch.
Lineare (6), Linear, leaves several times longer than wide, with parallel edges.
Linearifolia (2), linear-leaved.
Linearilobus, linear-lobed.
Lingula, generic name.
Liniflora, flax-flowered.
Linnæfolia, linnæa-leaved.
Linnæi, Linnæus (see Index).
Linoides, flax-like.
Lithocarya, stone-nut.
Lithospermoides, lithospermum-like, (see Index).
Litoralis (3), seashore,
Loasæfolia, loose-leaved.
Lobata, lobed.
Lobbi (4), Wm. Lobb, who collected in 1850-4.
Loculidical, dehiscence of compound pod (capsules) by splitting along the mid-rib (dorsal suture) of each carpel.
Longiflora (2), long-flowered.
Longifolia, long-leaved.
Longipes (3), long-peduncled.
Longipetala, long-petaled.
Lucida, shining.
Ludovicianus, Ludovicus.
Lugens, mourning.
Luisana, San Luis Obispo.
Lunate, shape of a new moon.
Lutea (5), **Luteolus** (2), yellow, yellowish.
Lutescens, yellow-growing.
Lyallii (3),
Lycopsoides (2), lycopus-like.
Lyoni (2), Wm. S. Lyon, Los Angeles.
Lyrata, *Lyrate*, lyre-shaped.

Macræi
Macrantha (2), large-flowered.
Macrocarpa (3), large-fruited.
Macrocera, large-horned.
Macrodon, large-toothed.
Macrophylla, large-leaved.
Macrothecum, large-anthered.

Macrosiphon, large-tubed.
Macrophyllum (3), large-leaved.
Macrostachya, large-spiked.
Maculata (2), spotted.
Major (3), larger.
Malachroides, like Malachra, another Mallows.
Malvæflora, mallows-flowered.
Malvæfolia, mallows-leaved.
Manzanita, little apple (The Spanish name).
Marah, genus name meaning bitter.
Maritima (11), maritime, sea-coast.
Marmoratus, shining-white.
Marshallii (3), C. C. Marshall, Humboldt Bay.
Marrubioides, horehound-like.
Marveanus, generic name.
Maximus, largest.
Media (2), medium.
Megacephalum, large-headed
Melanocarpa, black-fruited.
Melanthus, honey-flowered.
Millefoliata, thousand-leaved.
Mellita, honey-scented.
Membranacea, membranaceous.
Menziesii (17), Archibald Menzies, a Scotchman who collected in 1792-4.
Merous, in composition means parts; e. g., 5-merous parts in fives.
Mersinites, generic name.
Mertensiana (3) (see Index).
Meteloides, metel-like.
Mexicana (2), Mexican.
Michaeli (2), G. W. Michael.
Michneri,
Micrantha (10), small-flowered.
Microcarpus, small-podded.
Microcephalum, small-headed.
Microdon, small-toothed.
Micromeres, small (in all its parts).
Micropetala, small-petaled.
Microstachys, small-spiked.
Mildredæ, Mildred Holden daughter of the astronomer.
Millefolium, generic name meaning thousand-leaved.
Millegrana, a thousand grains.
Mineata, vermilion-red.
Minima (4), smallest.

Minor, smaller.
Minuta, minute.
Missouriensis, Missouri.
Mitracarpa, miter-podded.
Modesta, modest.
Monandrum, one-stamened.
Monantha (2), one-flowered.
Moniliformis, necklace-formed.
Monospermum, one-seeded.
Montana (7), mountain.
Molle (6), soft.
Montioides, montia-like.
Moschatus (2), musky.
Mucronate, Mucronulate, tipped with a sharp point.
Muhlenbergii, Dr. Henry Muhlenburg.
Multicaulis, many stemmed.
Multifida (3), much-dissected.
Multiflora (3), many-flowered.
Multinervis, many-nerved.
Multivalis, many-valved.
Muricata (2), *Muricate*, rough with short projections.
Muriculate, finely muricate.
Muirii, John Muir, geologist, naturalist and author, Alhambra Valley, Solano County.
Mutica, **Mutilum**, cut short.
Myriantha, many-flowered.
Myrinsites, an old name.
Myrtifolia, myrtle-leaved.
Myrtilloides, myrtle-like.
Myrtillus, generic name.

Nana (2), pigmy, dwarf.
Namatoides, nama-like.
Nasutus, large-nosed.
Nemaclada, thread-branched.
Nemoralis, sylvan, grove.
Nemorosa (2), leafy.
Nevadense (5), Sierra Nevaea, Nevada.
Neviniana, Rev. J. C. Nevin, Los Angeles.
Nervosa, nerved.
Newberryi, Dr. J. S. Newberry.
Nigellæformis, nigella-shaped.
Niger, (**nigra**) (3), black.
Nigrescens, blackening, becoming black.
Nitens, **Nitida** (3), shining.
Nivalis, **Nivea**, *snowy*.

Node, place on the stem where a leaf, a pair of leaves, or a whorl of leaves grow.
Nodiflora, node-flower (the peduncles from the nodes).
Nothofulvus, falsely-tawney (soon fading.
Nubigena, cloud-born.
Nuda, **Nudata** (2), naked.
Nudicaule, naked-stemmed.
Nudiflorus, naked-flowered.
Nummularia, money-wort.
Nutans, nodding.
Nuttaliana, **Nuttallii** (6), Thos. Nuttall, an eminent ornithologist and botanist, 1835.
Nutkanus (2), Nutka or Nootka.

Ob, the reverse ; e..g.,
Obcordate, cordate (leaf, petal, etc.) with the stem at the acute or rounded end of the blade, not in the sinus.
Obcordatus, obcordate.
Obispoensis, San Luis Obispo.
Obicularis, under Psoralea is a misprint for orbicularis.
Oblong, two or three times as long as broad, with parallel edges.
Oblonga, oblong.
Oblongifolia, oblong-leaved.
Obtuse, rounded at the end.
Obtusa, **Obtusata**, obtuse.
Obtusiflorum, obtuse-flowered.
Occidentalis (22), occidental, western.
Ocellata, eyed (i. e., a pair of spots), spotted.
Ochroleuca, *Ochroleucous*, yellowish-white.
Oculata, eyed.
Odorata, **Odoratissima**, sweet-scented and very sweet-scented.
Officinale (8), medicinal.
Oleracea, edible (the whole plant).
Oligosperma, few-seeded.
Oliganthum. few-flowered.
Olivaceum, olive-green.
Onustus, loaded, burdened.
Oocarpus, egg-pod.
Opulifolia, maple-leaved.
Opulus, maple.

Oppositifolia, opposite-leaved.
Orbicularis, *Orbicuar*, circular.
Orcuttiana, Orcuttii, C. R. Orcutt, naturalist, San Diego.
Oregana (11), **Oregonense** (2) Oregon.
Oreophylla, oreon-leaf.
Origanifolia, origanus-leaved.
Ornata (2), **Ornatissima,** ornate, most ornate.
Ornithorhynca, bird-beak.
Orthorhyncus, straight-beak.
Ovalifolia, oval-leaved.
Ovary, the part of a pistil containing ovules.
Ovata (5), *Ovate,* egg-shaped (used to describe flat organs. See Ovoid).
Ovatifolia, ovate-leaved.
Ovoid, egg-shaped, said of solids, (see *Ovate*).
Oxy, in composition means sharp or sour.
Oxycanthoides, like oxycanthus.
Oxycarpum, sharp-pod.
Oxycarya, sharp-nut.
Oxycoccus, sour-berry.
Oxynotus, markedly-sharp.
Oxypysus, sharp-bladder.

Pachylobus, thick-lobed.
Pachystachyus, thick-spiked.
Pacifica, Pacific Coast.
Palida, pale.
Palmata, *Palmate,* leaflets or lobes pointing away from the end of the petiole.
Palmeri (7). Dr. Edward Palmer, a noted collector of plants throughout the U. S. and Mex.
Palustre (9), swamp or marsh.
Panicle, a raceme of racemes or spikes.
Paniculata (2), *Paniculate,* bearing panicles.
Papillosus, *papillose.*
Papilionaceous, butterfly-like, like the flower of a pea.
Pardalinum, leopard.
Parishii, S. B. Parish, San Bernardino.
Parted, cut almost to the base or midrib.
Parryi (10), C. C. Parry, a noted field botanist and student of western plants.

Parviflora (19), small-flowered.
Parvifolia (7), small-leaved.
Parvula (2), very puny, small.
Parvum (2), small.
Pastoris, shepherd's.
Patagonica, Patagonia.
Patula, spreading, broad.
Pauciflora (4), few-flowered.
Paucisecta, few-lobed.
Pectinata, *Pectinate,* comb-like.
Pedata, *Pedate,* foot-like.
Pedicel, the stem of a small flower or a flower in a cluster.
Pedicellata, *Pedicellate,* flowers on pedicels.
Peduncularis (3), **Pedunculata** (2). *Pedunculate,* having a
Peduncle, stem of a flower.
Peltata (2), *Peltate,* shield-like.
Penduliflora, hanging-flowered.
Penicillata, *Penicillate,* having a brush-like appendage or tail.
Pentapetaloides, apparently five-petaled.
Peramænus, very-pleasing.
Peregrina, exotic, foreign.
Perenne (2), *Perennial,* having a root-stock.
Perfoliata (3), *Perfoliate,* the bases of opposite leaves united.
Perigynous, around the pistil on the receptacle.
Persistent, remaining longer than is common.
Petaloid, petal-like.
Petiolata, petioled, having a
Petiole, the stem of a leaf.
Petiolule, a little petiole.
Pharnaceoides, pharnaceon-like, like ginseng.
Physodes, bladder-like (the calyx).
Pickeringii, Dr. Chas. Pickering, Surgeon, Wilkes' Expedition, 1841.
Picta (2), painted, stained.
Pilosa (6) *Pilose,* covered with long distinct hairs.
Pilosissima, densely-pilose.
Pinetorum, of the pines, in pine forests.
Pinnata (2), *Pinnate,* leaflets on the sides of a common petiole (rachis).

Pinnatifida, *Pinnatifid*, deeply lobed in a pinnate manner.

Piperita, pepper-like.

Pisocarpa, pea-pod.

Placenta, the place where ovules grow.

Plantago, generic name.

Plattensis, Platt River.

Platycarpa, broad-podded.

Platycaule, broad-stemmed.

Platyloba, broad-lobed.

Platyphylla, broad-leaved.

Platysperma (2), broad-seeded.

Plebeia, low.

Plummeræ, Mrs. J. G. Lemmon (nee Plummer), an accomplished botanist.

Plumosa, Plumose, plume-like.

Pluriflora, many-flowered.

Portulacastrum, generic name.

Polycarpum (2), many-podded.

Polygaloides, polygala-like.

Polyphyllus (2), many-leaved.

Polysepalum, many-sepaled.

Polystachya (2); many-spiked.

Pome, a fruit like an apple.

Pomeridianum, afternoon.

Porrifolia, leek-leaved.

Posterior, away from the observer and next to the stem, upper.

Præmorsa, bitten into, eaten off.

Prenanthoides prenanthus-like.

Primuloides, primula-like.

Pringlei (2), C. G. Pringle.

Procera (2), tall.

Procumbens, *Procumbent*, trailing on the ground.

Prolifera (2), *Proliferous*, new branches or clusters of flowers arising from old ones.

Prostrata (4), *Prostrate*, lying flat on the ground.

Propinquus, much like another species.

Pruinosus, frost-like, with a white powder.

Pseudo-Capsicum, false-pepper.

Pubens, juicy, rapid growing.

Puberulus, *Puberulent*, covered with very fine short hairs.

Pubescent, covered with soft rather short hairs.

Pubescens (3), pubescent, becoming soft hairy.

Pudica, modest.

Pulchella (5), beautiful.

Pulegium, generic name.

Pulsiferæ (2), Mrs. Pulsifer Ames.

Pulverulenta, dusty, powdery.

Pumilla (6), diminutive, little.

Punctata, *Punctate*, dotted.

Pungens (3), *Pungent*, sharp.

Purpurascens (6), purple-growing.

Purpurea (4), purple.

Purshiana, **Purshii**, Frederick Pursh, (1800-20).

Pusilla (7) very small.

Pycnostachya, close-spiked.

Pycnantha, close-flowered.

Pygmæa (2), pygmy, dwarf.

Pyrolæfolia, pyrola-leaved.

Quadrangularis, four-angled.

Quadripetalum, four-petaled.

Quadrivales, four-valved.

Quadrivulnera, four-wounds ; i. e., with four red spots.

Quercetorum, of the oaks, growing among oaks.

Quercifolia, oak-leaved.

Raceme, a cluster of pedicillate flowers born along the main peduncle.

Racemosa (8), *Racemose*, flowers in racemes.

Rachis, the axis of a raceme or spike ; or of a pinnate leaf.

Radians, **Radiata**, radiating, radiate.

Radical, growing from a rootstock.

Ramosior, very *Ramose*, full of branches.

Ramosissimus (4), most ramose.

Ranunculifolia, ranunculus-leaved.

Raphanistrum, old generic name.

Rariflorum, rarely-flowering, few flowers.

Rattani (6), the compiler of this book.

Rawsoniana, Mrs. L. A. Peckenpah (nee Rawson).

Receptacle, the end of the peduncle upon which the organs of the flower grow.

Recurva, recurved.

Rediviva, living again, reviving after apparent death.

Reflexa, reflexed.

Reniformis, *Reniform*, kidney-shaped.

Repanda, *Repand,* wavy-margined.
Repens (2), **Reptans,** creeping.
Reticulated, netted-veined.
Retrorsa, *Retrorse,* bending back or point to the base as the lobes of some leaves.
Retuse, the apex slightly indented.
Revolute, rolled backward.
Rhodiola, generic name.
Rhombipetala, rhombus-petaled.
Rhomboides, rhombus-like.
Richardsoni
Rigida (7), rigid, stiff.
Rivale, brookside, growing along small streams.
Rivularis (3), rivulet-side, growing by rills.
Roezli.
Romanzofflana, Romanzovii, (see Index).
Rombifolia (2), rhombus-leaved.
Rootstock, the base of a perennial herb.
Rosaceus, rose-colored, rose-like.
Roscidum, dewy, moist.
Rosea (5), rose-colored.
Rotate, wheel-shaped.
Rothrockii, Dr. J. T. Rothrock, University of Penn.
Rotundifolia (8), round-leaved.
Rubellus (3), reddish.
Rubescens (2), reddening (with age).
Rubicunda, Ruber (Rubra, Rubrum), red.
Rufescens, slightly reddening in age.
Rugose, wrinkled.
Runcinate, leaves have coarse teeth or lobes pointing toward the base.

Sabinii.
Saccata, *Saccate,* having a sack-like cavity.
Sagittate, arrow-shaped.
Sambucifolia, elder-leaved.
Sanbournii.
Sanfordi.
Sanguinea (3), blood-red.
Santalanoides, santalanus-like.
Sarmentosa (2), *Sarmentose,* bearing slender twigs which coil around objects.

Sativa (3), cultivated.
Scabrella (2), *Scabrous,* rough, harsh.
Scape, a peduncle arising ' from the ground.
Scarious, dry and membranous.
Sceptrum, scepter or staff.
Scobarium, scurfy.
Scorpioid, coiled like a scorpion's tail.
Scouleri (5), Prof. Scouler, M. D., who collected before 1840.
Scripta, lined, scratched, written upon.
Scutellata, saucer-shaped.
Secunda (2), *Secund,* one-sided, bearing organs on one side.
Secundiflora, secund-flowering.
Semibarbata, half-bearded.
Sepium, of the hedges.
Septicidal dehiscence is the splitting of pods between the carpels.
Sericata (2), **Sericea** (3), *Sericeus,* silky.
Serpyllifolia, thyme-leaved.
Serpylloides, thyme-like.
Serrata, *Serrate,* with sharp teeth pointing toward the apex.
Serrulata (2), *Serrulate,* finely serrate.
Sessilis, *Sessile,* stemless.
Sessilifolia (2), sessile-leaved.
Setaceous, bearing bristles.
Setosa, bristly.
Shallon, an Indian name first applied to a genus.
Shastensis, Shasta.
Sheltonii (2), Rev. Shelton.
Siberica, Siberian.
Simplex (2), simple, plain.
Sinapistrum, generic name.
Sinuata, *Sinuate.*
Sinus, space between lobes.
Sitchensis (2), Sitka.
Smithii, B. H. Smith (?).
Soldanella, generic name.
Sonomense (3), Sonoma.
Sorediatus, granular, with rough spots.
Spadix, a fleshy spike of small flowers.
Sparsiflora (5) few-flowered.
Sparsifolium, few-leaved.
Spathe, the leaf enclosing a spadix.
Spathulifolium, spatulate-leaved.
Spathulata (2). *Spathulate, Spatulate,* near oblanceolate, with rounded apex

and blade abruptly narrowing below the center.

Spauldingii.

Speciosa (4), showy, beautiful.

Spectabilis (3), notable, remarkable.

Specularioides, specularia-like.

Spergulinum, spergu.a-like.

Spicata (5), *Spicate*, flowers in a *Spike*, a raceme with sessile or short-pediceled flowers.

Spinosus, spiny.

Spinulose, thorny.

Spiralis. coiling.

Spithamia (2), span-high.

Splendens (2), **Splendidum,** brilliant, shining.

Spur, a projecting appendage.

Spurium, bastard.

Squarrosa (2), *Squarrose*, spreading widely from the axis (leaves).

Stachyoides, stachys-like.

Staminea, thready, full of fibers.

Staminodia, stamens which have no anthers.

Standard, the upper petal of a papilionaceous flower.

Stanfordiana, Leland Stanford Jr.

Stellaris. *Stellate*, star-like.

Stenantha, short-flowered.

Stenocarpum, short-podded.

Stenoloba, short-lobed.

Stipe, stem of a pistil or of a pod above the receptacle.

Stipitata (2), *Stipitate*, having a stipe.

Stipels, the stipules of a leaflet.

Stipularis, *Stipulate*, bearing.

Stipules, a pair of appendages at the base of a petiole.

Stiveri, Dr. Chas. Stivers, San Francisco.

Stolonifera, *Stoloniferous*, bearing.

Stolons, branches which bending down take root.

Stramonium, generic name.

Striata, *Striate*, marked with grooves or channels.

Stricta (4), *Strict*, straight and narrow.

Strigosa, *Strigose*, beset with rigid scale-like appressed hairs.

Strigulosa, *Strigulose*, finely strigose.

Strobiliacea, Strobilina, cone-like, like the cone of a pine.

Suaveolens, sweet-scented.

Subacaulis, nearly acaulescent.

Subcordata (2), nearly cordate.

Subglochidiata (4), somewhat glochidiate.

Subinclusa, partly included.

Subpinnata, nearly pinnate.

Subpinnatifida, nearly pinnatifid.

Subspicata, somewhat spicate.

Subulata, *Subulate*, awl-bearing.

Subvestitum, partly covered.

Succulent. fleshy, juicy.

Suffrutescens, (2), *Suffrutescent*, shrubby at the base.

Suffruticosa, *Suffruticose*, shrubby.

Suksdorfii (2), Wm. Suksdorf.

Sulphureus, sulphur-colored.

Sylvatica (2), **Sylvester, Silvestris,** forest, of the woods.

Systyla, close-stemmed.

Tagitina, tagites-like, like the French marigold.

Tanacetifolia, tansy-leaved.

Taraxacifolia, dandelion-leaved.

Tatula, generic name.

Tenax (2), tenaceous, tough.

Tenella (7), **Tener,** tender, delicate.

Tenuiloba (2), narrow-lobed.

Tenuiflora (3), slender-flowered.

Tenuifolia, narrow-leaved.

Tenuis (2), slender.

Tenuissimus, very slender.

Terete, rounded cylindrically.

Ternata, *Ternate*, in threes.

Terrestris, low, on the ground.

Tessellata, tessellated, marked in squares (the nutlets).

Tetraphyllum (2), four-leaved.

Tetragona, four-sided.

Texana, Texas.

Thapsus, generic name.

Theophrasti, Theophrastus.

Thurberi, Dr. Geo. Thurber, whose special work was upon grasses.

Thyrse, a dense compound raceme.

Thyrsiflorus (3), thyrse-flowering.

Tinctoria, color-giving.

Tolmiei (2), Dr. Wm. F. Tolmie, Victoria, who came to the N. W. Coast in 1833.

Tomentella, woolly with fine matted hairs.

Tomentosa (5), *Tomentose,* covered with matted, woolly hairs.

Tortuosus, twisted.

Torulose, swollen at intervals.

Torus, an enlarged or broadened receptacle.

Torreyana, Torreyi (7), Dr. John Torrey, a chemist and the greatest American botanist of his time.

Trachyandra, rough-anther.

Trachycarpa, rough-pod or other fruit.

Tremuloides, quaking.

Tribracteum, three-bracted.

Trichantha, hairy-flowered.

Trichocalyx, hairy-calyx.

Trichocarpa, hairy-fruited.

Trichopodus, hairy-stemmed (the peduncle).

Tricolor (2), three-colored.

Tridentata (2), trident, three-toothed.

Trifida (4), three-cleft.

Triflorum, three-flowered.

Trifoliata, *Trifoliate, Trifoliolate,* bearing three leaflets.

Trilliifolia (2), trillium-leaved.

Triloba, three-lobed,

Triphylla (3), three-leaved.

Trixago, generic name.

Trolliifolium, trollius-leaved (one "i" omitted in text).

Truncata (3), *Truncate,* ending abruptly as though cut off.

Tuberosa (2), tuber-bearing.

Turbinate, top-shaped.

Ulignosa (2), juicy.

Umbel, a cluster of flowers having the pedicels all growing from the top of the peduncle.

Umbellata (6), umbel-bearing.

Unalaskensis (2), Unalaska.

Undulata, undulate.

Unguiculata, clawed.

Uniflora (6), one-flowered.

Unifolia, Unifoliata, one-leaved.

Unilateralis, one-sided.

Ursinus (2), bear.

Urticifolius, nettle-leaved.

Usatissimum, most-useful.

Utahensis, Utah.

Uva-ursi, generic name meaning bearberry.

Vaccaria (2), generic name.

Vagans, wandering.

Valerandi.

Validum, stout.

Variegata, variegated.

Variabilis, variable.

Variicolor, variously-colored.

Veitchianus, Veitch, an English gardener.

Velutina (3), velvety, fleecy.

Venenosus, poisonous.

Venusta (3), beautiful.

Verecunda, modest.

Verna (3), spring, early.

Versicolor, variable-colored.

Verticillaris, Verticillata, Verticillate, whorled.

Vernicosa, varnished.

Vesca, weak.

Vestita (3), clothed.

Victoris, Victor K. Chestnut.

Villosa (4), *Villous,* clothed with long hairs.

Viminea, willow-like, full of slender, osier-like branches.

Violacea, violet.

Virescens, green-growing, vigorous.

Virga, Virgata (3), *Virgate,* made up of slender shoots.

Virginiana (2), **Virginica** (3), **Virginiansis,** Virginia.

Viride (2), green.

Viridescens, green-growing,

Viscida, Viscidula, Viscosum, Viscosissimum, viscid, viscous, very sticky.

Vitifolius, grape-leaved.

Volubis, twining.

Vulgaris (3), Vulgatum, common.

Wallacei, Wm. A. Wallace.

Washingtonianum, Lady Washington.

Watsoni (2), Sereno Watson, author of "Botany of the King Exploration," "Botany of the California Geological Survey," etc.

Webberi (2), Dr. D. G. Webber of Sierra Co.

Weedii.

Wheeleri, Lieut. Geo. Wheeler, 1871–5.

Whipplei. Lieut. A. W. Whipple, 1853.

Whitlavia, a generic name.

Williamsoni, Lieut. R. S. Williamson, 1853.

Wings, the side petals of papelionaceous flowers.

Wormskjoldii.

Wrangelianus. Bar:. von Wrangel, Gov. of Russian possessions, lived at Bodega, 1829.

Wrightii, Chas. Wright.

Xanti, Xantiana, L. J. Xantus de Vesey, 1857-59.

Ziziphoroides, zizipus-like, like jujube.

INDEX AND GLOSSARY OF GENERIC NAMES.

.*. Names given in honor of men are followed by the names of the individuals thus honored. Local names are in italics. When such a name is not given in the text, the number or numbers of the species, or of the section, are given with the generic name in parentheses. *Baby Eyes*, for example, is a local name applied to the second and third species of *Nemophila*; and all the species in Section Two of *Calochortus* are called *Butterfly Lilies*.

The last page number refers to the *Key to Genera and Species;* the other number or numbers to the *Key to Orders*, where new species and corrections are found:

Abronia, delicate 69
Acæna 118
Acanthomintha, spiny-mint 172
Acer, sharp or strong 30, 104
Achlys 82
Aconite, Aconitum, the ancient Greek name 82
Actæa, elder-like 82
Adenostoma, glandular stoma (leaf openings) 117
Adolphia 102
Æsculus 30, 104
Agrimonia, prize of the field 118
Agrostemma,crown of the field 25
Alchemilla, the Arabic name 117
Alder 71
Alfalfa (Medicago, 1) 108
Alfilaria (Erodium) 101
Algaroba 32
Allionia 69
ALISMACEÆ, Alisma, water 72, 174
Allium, hot or burning 74, 180
Allocarya, all the nuts (maturing) 58
Allotropa, turning all ways (the flowers) 139
Alnus, near the river 71
Alsinanthemum, grove flower 50
Alsine, Alsinella, grove 25
Alyssum, allaying anger 86
Amelanchier, the French name 118
Ammannia, John Ammann, Russian (Misspelled in the text).. .... 38, 124
Amorpha, formless 35, 110
Ampelopsis, resembling a grape vine 29

Amsinckia, Wm. Amsinck, Hamburg........   ....57, 153
ANACARDIACEÆ...30, 104
Anagallis, cheering.   142
Androsace, a shield.. . 50
Anemone, wind..17, 79
Anemopsis, anemone-like.. 71
Antirrhinum, snout-like... 159
Aphyllon, leafless....   169
Aplectrum, spurless.. 176
APOCYNACEÆ, **Apocynum**, dog-bane..50, 142
Aquilegia, eagle (petals like eagle's claws)............. 81
Arabis, Arabian... 87
Aralia... 44
ARALIACEÆ.. 43
Arbutus, the Latin name... 137
Arctostaphylos, bear-berry..48, 137
Arenaria, sand-plant...........   95
Argemone, eye-cure..20, 83
ARISTOLOCHIACEÆ, **Aristolochia**.. 68
Armeria, Latin name of a similar plant....................................... 140
Aruncus.. 115
Asarum.. 68
ASCLEPIADACEÆ, **Asclepias**, .Esculapius, the Father of Medicine.............51, 142
Ash... 71
Astragalus 31, 111
Athysanus, without a fringe (otherwise like Thysanocarpus).................. 90
Audibertia, Audibert, a Frenchman...66, 173
Azalea (Rhododendron, 1, 2)...49, 139

Baby Eyes (Nemophila, 2, 3).. 150
Baneberry (Actæa)... 82
Barbarea, St. Barbara.....'....... 88
Barberry (Berberis).. 82
Bayberry... 71
Bedstraw (Galium).... ....46, 133
Behria, Dr. Behr (see note under Triteleia)................................. 182
Bellardia... 64
BERBERIDACEÆ...18, 82
Berberis.. 82
Bermudiana ....,................................. 73
Bergia 98
BETULACEÆ.. 71
Bilberry (Vaccinium, 1 to 7).. 137
Bigroot (Megarrhiza).. 129
Bindweed (Convolvulus, 6)... 156
Birch.. 71
Blackberry... 115
Bladderwort 169
Bleeding-heart. (Dicentra, 1)... 84
Bloomeria, H. G. Bloomer... 181

Blue-curls (Trichostema)... 171
Blue-eyed Grass (Sisyrinchium, 1)................................... 178
Blue-weed (Heliotropium).. 153
Boisduvalia, J. H. Bois Duval, a noted French naturalist of the 19th century, 39, 42, 128
Bolandra, Henry N. Bolander, a successful teacher and a noted botanist........ 120
Bolelia... 47
Borraginaceæ..57, 152
Borrago.. 57
Boschniakia, Boschniaki, a Russian................................. 169
Box-Elder 71
Boykinia, Dr. Boykin of Georgia.............................37, 120
Brasenia...18, 83
Brassica, Brassic, the Celtic name................................. 88
Brevoortia. J. C. Brevoort of New York............................ 183
Brodiæa, Jas. Brodie, a Scotch botanist.......................74, 181
Brossæa 49
Brunella, from the German name of a disease for which this plant was a remedy. 173
Bryanthus, moss-flower (growing in mossy places)................... 138
Buckbean (Menyanthes).. 105
Buckeye, Æsculus... 104
Buda... 25
Bur-clover 108
Burning-bush (Euonymus).. 102
Butneria... 36
Buttercup (Ranunculus)... 80
Butterfly Lily (Calochortus, § 2).................................. 186
Button-bush.. 133

Cactaceæ (Cactus) 43, 129
Calamintha, beautiful mint.. 172
Calandrinia, J. L. Calandrini, an Italian.....................26, 96
Calochortus, beautiful grass.................................76, 186
Caltha, cup 81
Calycanthaceæ, **Calycanthus,** calyx-flower......................36, 118
Calypso, the nymph Calypso.. 176
Calyptridium...26, 97
Camassia, the Indian name Camass.................................. 183
Campanulaceæ, **Campanula,** a bell..............................48, 135
Canaigre (Rumex)... 70
Capnorchis... 20
Capparidaceæ..23, 90
Caprifoliaceæ, **Caprifolium,** goat leaf............................. 131
Capsella, a little box.............................21 (figured), 23, 89
Cardamine, heart cure..........................21 (figured), 86
Carpenteria, Prof. Carpenter of Louisiana........................ 121
Carpet-weed (Mollugo).. 130
Caryophyllaceæ..24, 92
Cascara Sagrada, sacred bark....................................... 29
Cassiope, the goddess of that name............................... 138
Castilleia, D. Castillejo, a Spanish botanist.................64, 163

Catchfly (Silene).. 92
Caulanthus, stem-flower (petals stem-like)...................... 23, 88
Ceanothus, old Latin name... 102
CELASTRACEÆ..29, 102
Centunculus, unknown meaning... 142
Cephalanthera (the single anther, like a head, surpasses the stigma).......... 177
Cephalanthus, head-flower.. 133
Cerastium, horn plant (the pods like horns)....................... 94
Cerasus, name of the city in Western Asia from which cultivated varieties of
    cherries were first sent to Europe................................. 35
Cercis, shuttle (the pods like a weaver's shuttle)...................35, 113
Cercocarpus, tail-pod (see Fig. 79 in Exercises) 115
Cereus, wax or wax-like.. 130
Chamæbatia, Chamæbatiaria, from the Greek signifying near or on the
    ground (low or dwarf plants)...................................... 115
Chamisal, Chamise (Adenostoma)...................................... 117
Cheiranthus, hand-flower (?)......................................23, 88
Chelone, a tortoise or turtle (the flower resembling a turtle's head)............. 161
Cherry (Cerasus, p. 35, Prunus, 2, 3, 4, 5)......................... 114
Chia (Salvia, 2)... 173
Chickweed (Stellaria, 7)... 94
Chimaphila, winter-lover... 139
Chlorogalum, green milk (greenish white juice) 183
Chorizanthe 70
Chrysosplenium, golden spleen.................................... 121
Circæa, Circe the enchantress................................... 128
CISTACEÆ..23, 91
Cladothamnus, branching-bush..................................... 139
Clarkia, Gen. Wm. Clarke, who crossed the Rocky Mts. in 1803 with Meriwether
    Lewis, 39 (figured)...42, 128
Claytonia, Dr. John Clayton, an early botanist of Virginia. (In the figure, p.
    26, *a* is *C. exigua*; *b*, *C. perfoliata*)....................... 26, 97
Cleavers (Galium)...46, 133
Clematis, vine. 79
Cleome.. 90
Clintonia, Gov. De Witt Clinton of New York 188
Clover (Trifolium)..31, 33, 107
Cneoridium 101
Coffee Tree (Rhamnus) .:... 29
Collinsia, Zaccheus Collins, a botanist of Philadelphia............62, 160
Collomia, glue or gluten (seeds mucilaginous)..................... 53
Columbine (Aquilegia) 81
Comandra, hairy stamens (in the key)............................. 15
COMPOSITÆ 46
CONVOLVULACEÆ, Convolvulus, the old Latin name meaning a twiner.........64, 156
Coptis, from a Greek word meaning cut (leaves finely cut)............. 18, 81
Corallorhiza, coral-root... 176
Cordylanthus, club-flower 168
CORNACEÆ, Cornus, horn (wood horn-like)........................44, 131
Corn-spurry (Spergula)... 95

Corydalis, the old Greek name.. 84
Cotton.. 27
Cottonwood.. 71
Cotyledon, the old Greek name... 123
Crab-apple (Pirus)... 118
Cranesbill (Geranium).. 28, 101
CRASSULACEÆ 38, 122
Cratægus, strength 118
Cream-cups (Platystemon)...19 (figure), 20, 83
Cranberry (Vaccinium, 9).. 137
Cressa, Cretan woman.. 156
CRUCIFERÆ..21, 84
Cryptanthe, hidden flower.. 59
CUCURBITACEÆ, Cucurbita, Latin for gourd, the type of the order............43, 129
Currant (smooth-stemmed species of Ribes)...............................37, 122
Cuscuta, the Arabic name changed.. 156
Cypridium, Venus' Slipper... 177
Cycladenia, circle gland (a ring of glands around the pistil)............... 142
Cynoglossum, dog's tongue... 155

Damasonium.. 175
Darlingtonia, Dr. Wm. Darlington, a noted botanist of Pennsylvania.......... 83
DATISCACEÆ, Datisca...43, 129
Datura, the Arabic name modified.. 157
Death-Camass (Zygadenus, 2)... 188
Delphinium, dolphin (shape of flower)...................................17, 18, 81
Dendromecon, tree-poppy... 84
Dentaria, tooth (the tubers toothed).. 86
Dicentra, two spurs..20, 84
Dichelostemma 74
Dichondra, double mass (fruit double)....................................... 64
Digitalis, from the Latin for a thimble (the corolla fits the finger)....... 166
DIPSACACEÆ, Dipsacus..46, 134
Dirca (in the Key).. 16
Disporum 45
Distigia (Lonicera) two cloaks (the pair of bracts)......................... 75
Dodecatheon, twelve gods..50, 141
Dodder (Cuscuta).. 156
Dogwood (Cornus, 1)... 131
Downingia, A. J. Downing, author of several works on horticulture........47, 135
Draba, acrid (leaves)... 86
Draperia, John W. Draper, historian.. 150
DROSERACEÆ, Drosera, dewy (the leaves exude liquid)......................38, 123

Echinocactus, hedge-hog cactus.. 130
Echinospermum, hedge-hog seed...58, 155
ELATINACEÆ, Elatine, Greek name of the fir tree.........................26, 98
Elderberry (Sambucus)...45, 132
Ellisia, John Ellis, an English botanist................................54, 150
Emmenanthe, persistent flower... 152

Epilobium, violet on a pod................................39 (figured), 40, 125
Epipactis, the Greek name.. 177
ERICACEÆ...48, 136
Eriodictyon, woolly and veiny (the leaves)................................56, 152
Eriogonum, woolly joints... 70
Eriogynia, woolly pistil... 115
Erodium, heron (the fruit like a heron's bill).............................29, 101
Erysimum, a cure-all...23, 88
Erythræa, red... 144
Erythronium, red (inappropriate)......................................76, 185
Eschscholtzia, J. F. Eschscholtz, a German naturalist, who visited San Fran-
    cisco, San Jose, and Monterey in October, 1816..............19 (figured), 20, 84
Eucharidium...42, 128
Eulobus, truly a pod (pod 3 or 4 inches long)............................41, 126
Eunanus, very charming... 63
Euonymus, good name.. 102
Evening Primrose, (Œnothera, 1) 126
Fatsia.... 44
FICOIDEÆ.. 43, 130
Filaria or *Filaree* (Erodium).. 101
Flax, (Linum)... 100
Flœrkia, Flœrke, a German botanist...................................... 29
Forgetmenot, (Myosotis)... 153
Foxglove... 166
Fragaria, fragrant (the fruit)... 116
FRANKENIACEÆ, **Frankenia**...29, 92
Frasera, John Fraser, an English botanist............................... 144
Fraxinus... 142
Fremontia, Gen. John C. Fremont......................................28, 100
Fringe-pod (Thysanocarpus)......................................21 (figured), 90
Fritillaria, checkered..75, 185
FUMARIACEÆ...20, 84

Galium..46, 133
Garrya... 131
Gaultheria (Gaulthier), a French physician at Quebec..................... 138
Gayophytum...40, 126
GENTIANACEÆ, **Gentiana**...51, 143, 144
GERANIACEÆ, **Geranium,** crane (fruit like the bill of a crane)..............28, 101
Geum, good tasting... 116
Gilia, Philip Gil or Gilio, a Spanish botanist..........................52, 145
Githopsis, like Githago... 48, 135
Glaux, sea-green.. 141
Glycyrrhiza, sweet root.. 110
Godetia, 39 (figured)...42, 127
Goldthread (Coptis)...18, 81
Gomphocarpus, club fruit or peg-pod..................................51, 143
Goodyears, John Goodyear.. 177
Gooseberry (the prickly Ribes).......................................37, 122
Goosefoot (Potentilla, 2).. 116

Grape..29, 102
Gratiola, herb of grace.. 166
Ground Cherry, (Physalis)... 138

Habenaria, thong (the spur)..72, 176
HALORAGEÆ..38, 124
Hastingsia, Judge C. Hastings of San Francisco, who assisted in the publication
   of the State Botanical Survey Report of Cal.... 183
Helianthemum, sun flower.. 91
Heliotropium, sun-turning... 153
Hemitotes (Newberrya)... 49
Hernaria.. 25
Herpestis, a creeper... 166
Hesperochiron, the western Chiron.. 152
Hesperoscordum... 75
Heterodraba, other or false Draba.. 90
Heterocodon, bells (flowers) differing 136
Heterogaura, other or false Gaura.. 128
Heteromeles.. 118
Heuchera, Dr. H. Heucher 121
Hibiscus... 100
Hippocastenum, horse-chestnut.. 30
Hippuris, mare's tail.. 124
Holodiscus, all disk (the flower)....................................35, 115
Honeysuckle (Lonicera) 45, 132
Hookera, Sir Josoph Hooker, a noted English botanist.................74, 182
Horehound.. 173
Horkelia...35, 117
Hosackia, Dr. David Hosack, a Philadelphia botanist........... 31 (figured), 34, 108
Howellia, Joseph and Thos. T. Howell, Oregon, botanists................... 48
Huckleberry (Vaccinium, 8)... 136
HYDROPHYLLACEÆ, **Hydrophyllum**, water leaf.............................54, 149
HYPERICACEÆ, **Hypericum**...25, 98

ILLECEBRACEÆ...25, 96
Ilysanthes, mud flower... 166
Indian Lettuce (Claytonia perfoliata)...................................... 97
IRIDACEÆ, **Iris**, Iris, goddess of the rainbow.........................73, 178
Isomeris, equal parts.. 90
Isopyrum, equal wheat.. 81
Ivesia...35, 117

Jussiæa, A. J. Jussieu, a noted French botanist......................39, 125

Kalmia, Peter Kalm, a Swede.. 138
Kelloggia, Dr. Albert Kellogg, of San Francisco, who was a lifelong student of
   our Western plants... 133
Knotgrass.. 70
Koelia.. 66

Krameria, Kramer (two German brothers)................................... 92
Krynitzkia, Prof. J. Krynitzki of Cracow...........................58, 154

LABIATÆ...65, 170
Lace-pod (Thysanocarpus, 1)............................21 (figured), 90
Lady's Slipper (Cypripedium).. 177
Larkspur (Delphinium)... 81
Lathyrus.. 112
LAURACEÆ (*Laurel*)... 71
Laurentia, M. A. Laurenti of Bologna (18th Century)................... 135
Lavatera, Lavater (two brothers, Zurich)............................... 99
Leduni, the old German name... 139
LEGUMINOSÆ...30, 104
Lemmonia, Prof. J. G. Lemmon, a noted botanist, author of a work on West
    Coast Coniferæ, etc... 56
LENNOACEÆ..49, 140
LENTIBULARIACEÆ..65, 169
Lepidium, a little scale (pods)...................................21, 90
Lepigonum, scaly joint...25, 95
Lesquerella, Leo Lesquereux, a noted bryologist...................... 2 ?
Leucocrinum, white lily.. 183
Leucothoe, a mythical goddess (see **Lucothoe**).................... 138
Lewisia, Capt. Lewis who crossed the continent with Clarke in 1803-6 . 98
LILIACEÆ...73, 178
Lilium, *Lily*.. 184
Limnanthes, marsh or mud flower..................................29, 101
Limodorum, mud lover.. 73
Limonium, marsh-wort.. 49
Limosella, little mud plant... 166
LINACEÆ...28, 100
Linanthus, flax-flower (like flax).................................. 52
Linaria, flax-like..61, 159
Linnæa, Carl von Linnæus, a Swede, the first great systematic naturalist....... 132
Linum, thread (used to make thread)..............................28, 100
Lippia, Aug. Lippe, a Frenchman..................................60, 170
Liquorice, (Glycyrrhiza).. 109
Listera, Dr. Martin Lister, an Englishman........................... 177
Lithospermum, stone-seed..57, 153
LOASACEÆ..42, 128
LOBELIACEÆ, **Lobelia**, Mathew Lobel, physician to James I.........41, 134
Loeflingia, Loefling, a botanist of the 18th Century................ 96
Lonicera, Adam Lonicer, a German.................................45, 132
Lophanthus, crest-flower.. 173
Lotus, Latin name for one of its species............................ 34
Lucern (Medicago, 1).. 108
Lucothoe, a misprint for Leucothœ................................... 138
Ludwigia, Prof. C. D. Ludwig, Leipsic...........................39, 125
Lupinus, *Lupine*, wolf...32, 105
Lychnis, a lamp...25, 94

Lycium, the Greek name 157
Lycopus, wolf-foot... 171
Lysimachia, peacemaker.... 141
LYTHRACEÆ, Lythrum..38, 124

Madrona, the Mexican name of our Arbutus............................... 137
Maianthemum, May-flower..75, 184
Mallows, (Malva)............................ ....27, 99
Malus, the Latin name for apple... 35
MALVACEÆ, Malva, soft...27, 98, 99
Malvastrum, like mallows................................... ....27, 99
Malveopsis, mallows-like.. 28
Mammillaria.. 130
Manzanita, little apple.. 137
Maple, (Acer)... 104
Marrubium, bitter-juice.. 173
Marsh-mallow.. 27
Meadia 50
Meconopsis, poppy-like, 19 (figured)...............................20, 84
Megarrhiza, big-root 129
Melilotus, honey lotus.. 108
Melissa, bee plant.. 66
Mentha, Minthe, a mythical personage.... ....65, 171
Mentzelia, C. Mentzel of Brandenburg...............................42, 128
Menyanthes, moon-flower... 145
Menziesia, Archibald Menzies, who with Vancouver visited the Pacific Coast in
    1791–5.. 138
Mertensia, F. C. Mertens of Bremen.................................... 153
Mesembryanthemum, mid-day flower...................................... 130
Mesquit, the Mexican name... 32
Micrampelis, small vine... 129
Microcala, a little beauty.. 143
Micromeria, small.... ....66, 172
Mimetanthe, monkey flower... 63
MIMOSÆ, mimio..... 30
Mimulus, ape..63, 163
Mirabilis, wonderful.. 69
Mitella, little miter (the fruit)..................................37, 121
Modiola, little cup... 28
Mohavea, Mohave... 160
Mollugo.. 130
Monardella, little Monarda, Dr. N. Monardes... ....65, 171
Moneses, only one (flower).. 139
Monk's Hood... 82
Monotropa, turned to one side... 140
Montia..26, 97
Morning Glory (Convolvulus)... 155
Mountain Balm (Eriodictyon)... 150
Muilla, anagram of Allium... 181
Mullein (Verbascum).:... 138

Mustard, Brassica.. 87
Myosotis, mouse ear.. 153
Myosurus, mouse tail... 80
MYRICACEÆ.. 71
Myrica, flowing (grows by rivers)................................ 70
Myriophyllum, a thousand leaves................................. 124

Nama, a spring of water.....................................56, 152
Narthecium, rod or wand... 189
Nasturtium, nose twister (pungent odor)...................23, 89
Navarritia.. 52
Negundo...50, 104
Nemacladus, thread branches.................................... 134
Nemophila, grove lover....................................54, 150
Nepeta, Nepet, a Tuscan town.................................... 66
Newberrya, Prof. J. S. Newberry, a noted geologist and student of extinct
 .. plants..49, 140
Nicotiana, John Nicot, who introduced tobacco into France.......... 158
Nine-Bark, Physocarpus.......................................35, 115
Nuphar, Arabic name of Water Lily..........................18, 83
Nuttallia, Thos. Nuttall, ornithologist and botanist.........35, 114
NYCTAGINACEÆ... 69
NYMPHÆACEÆ, **Nymphæ**, water nymph.............................18, 82

Oak... 71
Œnothera, thirst maker.....................................41, 126
Odontostomum, toothed mouth.................................... 184
OLEACEÆ..50, 142
Omphalodes, navel-like... 155
ONAGRACEÆ..39, 124
Onion (Allium).. 180
Opulaster.. 35
Opuntia, Opuntii, a region in Greece........................... 130
Orchiastrum.. 73
ORCHIDACEÆ...72, 175
Oregon Crab Apple (Pirus).. 118
Oregon Grape (Berberis, 2)....................................... 82
OROBANCHACEÆ...65, 169
Orthocarpus, straight pod...................................64, 167
Osmaronia.. 35
Oso-Berry (Nuttallia).. 114
Oxalis, acid...29, 101
Oxys, acid... 29
Oxytheca, sharp anther... 70

Pachystima... 102
Pæonia, Dr. Pæon... 82
Palmerella, Dr. Edward Palmer, a field botanist, who has collected extensively
 in the United States and Mexico.................................. 135

Pansy (Viola)... 91
PAPAVERACEÆ, **papaver**, thick milk.............................19, 20, 83
Parnassia. Mt. Parnassus.. 121
Paronychia.. 25
Pectocarya, comb-nut.. 156
Pedicularis, a louse.. 168
Pentacæna 96
Pentstemon, five stamens...62, 161
Peony, Pæouia... 82
Petunia, from Petun, the Brazilian name for tobacco, a similar plant........... 158
Phacelia, a bundle (flowers)..55, 150
Phænicaulis, exposed stems... 88
Philadelphus, a friendly brotherhood (of the stamens?)................... 121
Philbertia, J. C. Philbert, a French teacher of botany................... 143
Phlox, a flame (the flowers)...53, 145
Pholisma, scale (scaly stem)... 140
Photinia (Heteromeles, *Toyon*)... 118
Physalis, bladder (the inflated fruiting calyx).......................... 138
Physocarpus, bladder pod... 115
Pickeringia, Dr. Chas. Pickering, who botanized from the Columbia by way of
    Sacramento Valley to San Francisco in 1841..........................35, 105
Pin-clover (Erodium)... 101
Pimpernel (Anagallis).. 142
PIPIRACEÆ.. 71
Pipe Vine (Aristolochia)... 68
Pipsissewa (Chimaphila).. 139
Pirus, old name of the pear... 118
Plagiobothrys, side cavity (not appropriate)............................ 157
PLANTAGINACEÆ, **Plantago**, sole of the foot (the common species growing on
    tramped ground...67, 174
Plantain (Plantago).. 174
Platanus. broad or ample (leaves and branches).......................... 71
Platyspermum, flat seed... 86
Platystemon, flat stamen.................................19 (figured), 20, 83
Platystigma, flat stigma.................................19 (figured), 20, 83
Plectritis (Valerianella. Suksdorf puts the plants of this genus in four genera,
    and makes several new species).... 134
Pleuricospora, seed at the side (of the pod)............................. 140
Plum (Prunus, 1)... 114
PLUMBAGINACEÆ...49, 140
Pogogyne, bearded pistil (style)... 172
Poison Oak (Rhus, 1)...30, 104
POLEMONIACEÆ..51, 145
Polemonium, the old Greek name.. 149
Polycarpon, many pods..25, 95
Polygala, much milk (stimulating secretion of in animals)................ 92
POLYGALACEÆ...24, 91
POLYGONACEÆ, **Polygonum**, many joints..................................69, 70
Poplar (Populus, tree of the people)....................................... 71
Poppy (Papaver).. 20

PORTULACACEÆ, **Portulaca**, juice bearer ....26,　96
Potentilla, potent (in disease)........　...35,　169
Poterium, cup.....　...　118
Prickly Poppy (Argemone)......　......................................19 (figured),　83
PRIMULACEÆ, **Primula**, first (to bloom)...49,　141
Prince's Pine (Chimaphila)...　139
Prosartes, to suspend (flowers pendulous)..75,　187
Prosopis...　.....................　32
Prunus, Latin name of plum...35,　114
Psoralea, scurfy, glandular........　...　110
Pterospora, winged seed...　140
Putty-root (Aplectrum)...　176
Pycnanthemum, dense flowered......................　...............................66,　171
Pyrola, pirus, a pear (leaves like)..　139

Quaking Ash...　71

Radish (Raphanus)..21 (figured),　90
Ramona........　...　66
RANUNCULACEÆ, **Ranunculus**, little frog.......................................17, 79,　80
Raphanus, quick coming (coming up early)...　90
Raspberry (Rubus, 2)...　115
Rattle-weed (Astragalus, the species with bladder-like pods, 7 to 19).............　111
RHAMNACEÆ, **Rhamnus**...............　...29,　102
Rhododendron, rose tree...49,　139
Rhus, red (the fruit of some species)...30,　104
Ribes..37,　122
Romanzoffia, Nicholas Romanzoff, a Russian nobleman who early in the 19th
　　century sent Kotzebue to this coast, accompanied by the naturalists
　　Chamisso and Eschscholtz...　152
Romneya..　83
Roripa...　23
ROSACEÆ, **Rosa**, *Rose*, red...35, 36, 113,　118
Rotala...38,　124
RUBIACEÆ...45,　133
Rubus, red (the fruit of some species)...　115
Rumex...　70
RUTACEÆ...29,　101

Saccaline..　70
Sage (Audibertia. Garden sage is a Salvia)..　173
Sagina, fatness...　95
Sagittaria, arrow bearer (the leaf)...　175
Salal (Gaultheria, 1)...　138
SALICACEÆ...　71
Salmon Berry (Rubus, 3)...　115
Salvia, saving, preserving..66,　173
Sambucus, name of musical instrument made of elder.....　...................45,　132
Samolus, pig's food...　142

Sand-Spurry (Lepigonum).. 95
Sand Verbena (Abronia).. 69
Sandwort (Arenaria)... 95
SAPINDACEÆ..30, 103
Saponaria, soap..25, 94
SARRACENIACEÆ..19, 83
Sarcodes, flesh-like... 140
SAXIFRAGACEÆ, Saxifraga, Saxifrage, stone breaker....................36, 37, 119
Scniznotus, cleft-back (the hoods of the flower split down the back)..........51, 143
Scoliopus, worm-stem (the scapes)... 124
Screw-pod Mesquit... 32
SCROPHULARIACEÆ 60, 158
Scrophularia, scrofula cure... 161
Scutellaria, little helmet.. 173
Sedum, sitting (habit of the plant)....................................... 123
Self-heal (Brunella).. 173
Senebiera, C. D. Senebier, Geneva... 90
Service-berry (Amelanchier)... 118
Sesuvium.. 130
Sheep-sorrel (Rumex acetosella)... 70
Shepherd's purse (Capsella, 2)....................................21 (figured). 89
Sherardia... 46
Shooting Star (Dodecatheon)...50, 141
Sibaldia, Robt. Sibald, Edinburgh.....................................35, 117
Sida.. 100
Sidalcea..27, 99
Silene, saliva (exudation from some species)...........................24, 92
Silkweed (Asclepiadaceæ, all our species)................................. 142
Sisymbrium 89
Sisyrinchium, pig's snout (the spathe)................................73, 178
Skullcap (Scutellaria).. 173
Skunk-weed (Gilia, the fetid species)................................52, 145
Smilacina, like Smilax (but not the so called Smilax of our gardens...........75, 184
Smilax, a rare woody climber with small green flowers in umbels (inadvertently
    omitted)..
Snow-berry (Symphoricarpos)..45, 132
Snow-plant (Sarcodes)... 140
Soap-root (Chlorogalum)... 183
SOLANACEÆ, Solanum, the Latin name....................................60, 157
Solonoa, Solano Co.. 51
Sorrel, (Oxalis).. 101
Specularia, mirror.. 135
Spergula, scatter... 95
Sphacele, Greek name of garden Sage....................................... 172
Spikenard... 44
Spinach... 43
Spiræa, to wind (in wreaths).. 114
Spiranthes, spiral of flowers...73, 177
Spraguea (Calyptridium) Sprague, a botanical artist, who illustrated Gray's
    Botanies)..26, 97

Stachys, a spike (the flowers in)..66, 172
Stanfordia, Gov. Leland Stanford... 89
Stanleya... 86
Staphylea, cluster... 104
Star-Flower (Trientalis)... 141
Statice, astringent...49, 140
Stellaria, star-like... 94
Stenanthium, narrow-flower... 189
STERCUDIACEÆ, bad odor, fetid...28, 100
Stonecrop (Sedum).. 123
Stramonium (Datura).. 167
Strawberry (Fragaria).. 166
Streptanthus, twisted flower (the petals)..................................22, 87
Streptopus, twisted stalk.. 187
Stropholirion, (Brodiæa 1), twining lily..................................... 181
STYRACACEÆ, Styrax..50, 142
Subularia, owl (pods owl-like)... 89
Suksdorfia, Wm. Suksdorf, State of Washington................................ 120
Sulivantia, Wm. S. Sullivant, a noted American bryologist (student of Mosses).. 120
Sundial (Lupinus).. 105
Sweet-clover (Melilotus)... 108
Sweet-scented Shrub (Calycanthus).. 118
Sycamore... 71
Symphoricarpos, fruit accumulator...45, 132
Synthyris.. 166

Tare (Vicia sativa).. 112
Tellima, anagram of Mitella...37, 120
Tetragonia, four-angled.. 43
Thalictrum, green-growing...17, 80
Thelypodium...22, 88
Thermopsis, Lupine-like.. 105
Thimble-berry (Rubus, 1)... 115
Thistle Sage (Salvia, 1)... 173
Thlaspi.. 89
Thrift (Armeria)... 140
Thysanocarpus, fringe-pod...21. 90
Tiarella... 121
Tiger Lily (Lilium pardalinum, which means Leopard Lily, a better name)...... 184
Tillæa, M. A. Tilli, an Italian...38, 122
Tissa.. 25
Toad Flax (Linnaria)... 150
Tolfieldia... 189
Tolmiea.. 120
Tonella, probably meaningless...62, 158
Toyon (Heteromeles), pronounce both o's long................................. 117
Trautvetteria, named for a German botanist................................... 80
Trichostema, hair stamens.. 171
Trientalis, three inches high...50, 141
Trifolium, three-leaf...31, 33, 107

Trillium, parts in threes 188
Triodanis.. 40
Triteleia 74, 182
Tropidocarpum ....21 (figured), 22, 89
Twayblade (Listera).... 177

UMBELLIFERÆ.. 44
Umbellularia 71
Unifolium. one leaf... 75
Utricularia, little bladders.. 169

Vaccaria... 25
Vaccinium.. 137
Vagnera... 75
VALERIANACEÆ, **Valeriana, Valerianella,** *Valerian,* King Valerius.....46, 133, 134
Vancouveria, Capt. Geo. Vancouver, who explored the Pacific Coast in 1792-5..18, 82
Veratrum, true black.. 188
Verbascum, beard.. 159
VERBENACEÆ, **Verbena**, old Celtic name modified........................... 67, 170
Veronica, St. Veronica (?) 166
Vesicaria, blister (pods inflated)....................................23, 89
Vibernum, tie (used to make withes).................................. 132
Vicia, bind...34, 112
Vine-Maple (Acer, 2).. 104
VIOLACEÆ, **Viola**, *Violet*, the old Latin name 23, 91
Virgin's Bower (Clematis) 79
VITACEÆ, **Vitis**, the best..29, 103

Wake-Robin (Trillium).. 188
Wall-flower (Cheiranthus)..23, 88
Walnut 71
Water-Plantain (Alisma).. 175
Water-Lily (Nuphar)...18, 83
Water-Shield (Brasenia)...18, 83
Whipplea, Lieut. A. W. Whipple, who was in command of a Government Survey
    Party, on the Pacific Coast in 1853-54................................ 121
Willow 72
Willow-herb (Epilobium)... 125
Wood Anemone (Anemone, 3).. 79

Xylothermia.. 35
Xerophyllum, dry leaf.. 189

Yerba Buena (Micromeria, 1).. 172
Yerba Mansa (Anemopsis)... 71
Yerba Santa (Eriodictyon)..56, 152
Yucca... 184

Zauschneria, M. Zauschner, a Bohemian botanist......................39. 125
Zizyphus.. 102
Zygadenus, yoked glands (on the petals)............................ 188